Engineering Communication

Engineering Communication

Second Edition

HILLARY HART
Department of Civil Engineering
University of Texas at Austin

PEARSON
Prentice
Hall

Upper Saddle River, New Jersey 07458

Library of Congress Cataloging-in-Publication Data on file

Editorial Director, ECS: *Marcia J. Horton*
Senior Editor: *Holly Stark*
Associate Editor: *Dee Bernhard*
Editorial Assistant: *Jennifer Lonschein*
Senior Managing Editor: *Scott Disanno*
Production Editor: *James Buckley*
Art Director: *Kenny Beck*
Cover Designer: *Bruce Kenselaar*
Art Editor: *Greg Dulles*
Manufacturing Manager: *Alan Fischer*
Manufacturing Buyer: *Lisa McDowell*
Marketing Manager: *Tim Galligan*

© 2009, 2005 Pearson Education, Inc.
Pearson Prentice Hall
Pearson Education, Inc.
Upper Saddle River, NJ 07458

The author and publisher of this book have used their best efforts in preparing this book. These efforts include the development, research, and testing of the theories and programs to determine their effectiveness. The author and publisher make no warranty of any kind, expressed or implied, with regard to these programs or the documentation contained in this book. The author and publisher shall not be liable in any event for incidental or consequential damages in connection with, or arising out of, the furnishing, performance, or use of these programs.

Printed in the United States of America
10 9 8 7 6 5 4 3 2 1

ISBN-10: 0-13-604420-4
ISBN-13: 978-0-13-604420-8

Pearson Education Ltd., London
Pearson Education Australia Pty. Ltd., Sydney
Pearson Education Singapore, Pte. Ltd.
Pearson Education North Asia Ltd., Hong Kong
Pearson Education Canada, Inc., Toronto
Pearson Educación de Mexico, S.A. de C.V.
Pearson Education—Japan, Tokyo
Pearson Education Malaysia, Pte. Ltd.
Pearson Education, Inc., Upper Saddle River, New Jersey

Contents

6 • REVISING: WHEN WILL I EVER BE FINISHED? 94

7 • SPEAKING: DO I REALLY HAVE TO STAND UP AND TALK IN FRONT OF ALL THOSE PEOPLE? 115

8 • PRODUCING ENGINEERING DOCUMENTS: THE FINAL PRODUCT 142

ESource Reviewers

We would like to thank everyone who helped us with or has reviewed texts in this series.

Naeem Abdurrahman, *University of Texas, Austin*
Stephen Allan, *Utah State University*
Anil Bajaj, *Purdue University*
Grant Baker, *University of Alaska–Anchorage*
William Beckwith, *Clemson University*
Haym Benaroya, *Rutgers University*
John Biddle, *California State Polytechnic University*
Donald Blackmon, *The Citadel*
Tom Bledsaw, *ITT Technical Institute*
Fred Boadu, *Duke University*
Jerald Brevick, *The Ohio State University*
Tom Bryson, *University of Missouri, Rolla*
Ramzi Bualuan, *University of Notre Dame*
Dan Budny, *Purdue University*
Betty Burr, *University of Houston*
Fernando Cadena, *New Mexico State University*
Joel Cahoon, *Montana State University*
Dale Calkins, *University of Washington*
Linda Chattin, *Arizona State University*
Harish Cherukuri, *University of North Carolina–Charlotte*
Arthur Clausing, *University of Illinois*
Barry Crittendon, *Virginia Polytechnic and State University*
Donald Dabdub, *University of CA–Irvine*
Richard Davis, *University of Minnesota Duluth*
Kurt DeGoede, *Elizabethtown College*
John Demel, *Ohio State University*
James Devine, *University of South Florida*
Heidi A. Diefes-Dux, *Purdue University*
Jerry Dunn, *Texas Tech University*
Ron Eaglin, *University of Central Florida*
Dale Elifrits, *University of Missouri, Rolla*
Tim Ellis, *Iowa State University*
Christopher Fields, *Drexel University*
Patrick Fitzhorn, *Colorado State University*
Julie Dyke Ford, *New Mexico Tech*
Susan Freeman, *Northeastern University*
Howard M. Fulmer, *Villanova University*
Frank Gerlitz, *Washtenaw Community College*
John Glover, *University of Houston*
John Graham, *University of North Carolina–Charlotte*

Laura Grossenbacher, *University of Wisconsin Madison*
Ashish Gupta, *SUNY at Buffalo*
Otto Gygax, *Oregon State University*
Malcom Heimer, *Florida International University*
Donald Herling, *Oregon State University*
Thomas Hill, *SUNY at Buffalo*
A. S. Hodel, *Auburn University*
Kathryn Holliday-Darr, *Penn State U Behrend College, Erie*
Tom Horton, *University of Virginia*
James N. Jensen, *SUNY at Buffalo*
Mary Johnson, *Texas A&M Commerce*
Vern Johnson, *University of Arizona*
Jean C. Malzahn Kampe, *Virginia Polytechnic Institute and State University*
Moses Karakouzian, *University of Nevada Las Vegas*
Autar Kaw, *University of South Florida*
Kathleen Kitto, *Western Washington University*
Kenneth Klika, *University of Akron*
Harold Knickle, *University of Rhode Island*
Terry L. Kohutek, *Texas A&M University*
Bill Leahy, *Georgia Institute of Technology*
John Lumkes, *Purdue University*
Mary C. Lynch, *University of Florida*
Melvin J. Maron, *University of Louisville*
F. Scott Miller, *University of Missouri Rolla*
James Mitchell, *Drexel University*
Robert Montgomery, *Purdue University*
Nikos Mourtos, *San Jose State University*
Mark Nagurka, *Marquette University*
Romarathnam Narasimhan, *University of Miami*
Shahnam Navee, *Georgia Southern University*
James D. Nelson, *Louisiana Tech University*
Soronadi Nnaji, *Florida A&M University*
Sheila O'Connor, *Wichita State University*
Matt Ohland, *Clemson University*
Kevin Passino, *Ohio State University*
Ted Pawlicki, *University of Rochester*
Ernesto Penado, *Northern Arizona University*
Michael Peshkin, *Northwestern University*
Ralph Pike, *Louisiana State University*
Dr. John Ray, *University of Memphis*

Engineering Communication

Engineering and Communication

1.1 I AM AN ENGINEER, NOT A PROFESSIONAL COMMUNICATOR

Maybe nobody has told you yet, but the truth is this: To be an engineer is to be a technical communicator. Engineering is a problem-solving profession, and clear communication leads to effective solutions. As an engineer, you will solve problems for people (often, large groups of people). Whether your specialty is civil, mechanical, electrical, chemical, biomedical, computer, aerospace, or any other engineering discipline, you will develop products, services, and environments needed by some segment of the public. In order to produce those goods and services, you will work collaboratively with dozens of other engineers, experts, and support personnel, designing, analyzing, refining, and building solutions. You, your collaborators, and your clients will have to understand exactly what each of you is doing every step of the way. That is why communication skills are critical to engineering work.

Communication means writing, speaking, and showing visuals to other people, as well as listening to and observing others and reading their work. Of course, most of us think of writing as the most difficult skill to master (although, for some of us, speaking in front of a crowd seems not only difficult, but also terrifying). Yet, writing well is critical for communicating your ideas and solutions. Let us say you have designed a new type of bus shelter for a large city in a rainy part of the United States. Your design is rendered very professionally, either in CAD or as an artist's rendering, and you have submitted the design to the municipal transportation authority. In addition to presenting the design (see Figure 1.1), you will want to explain in a proposal the main features of the design and the criteria you used to develop it, such as the following:

- Keeps out rain effectively
- Allows enough openness to discourage theft and crime
- Trash does not collect inside

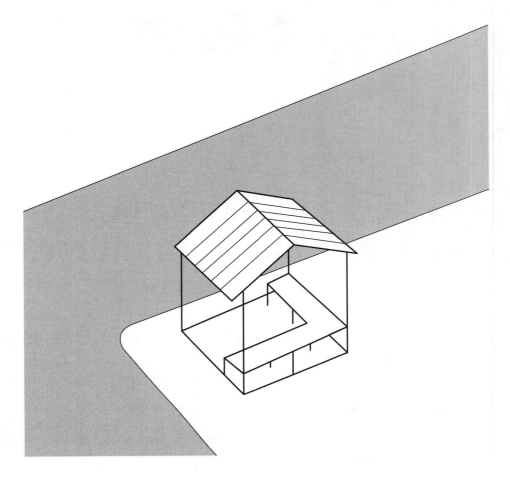

Figure 1.1
Artist's rendering of proposed bus-shelter design.

You will use pictures and words to argue that this design should be accepted and built. You may also include a line graph displaying the drop in bus-shelter crime experienced by another city that adopted a similar design. But your design improves on the previous one by being less costly: It can be assembled away from the site and installed on-site within a matter of hours. Sounds good . . . but only as good as you make it sound with your words and make it look with your visuals.

Communication is a part of almost every phase of engineering work. This chapter presents some basic tenets to remember throughout your career.

1.2 WRITING DEMONSTRATES MY COMPETENCE AS AN ENGINEER

Someone can point to a proposal you write for, as an example, a new type of bus shelter and say, "See what a good solution this is; see this clear explanation of how this design solves the problems." Your competence as an architectural engineer is proven by that proposal. The same concept is true for all the engineering disciplines. Inside your own organization as well, writing demonstrates that you understand the work you do and the jobs of the people with whom you work. Even your informal writing, such as memos and e-mail messages, demonstrates this understanding. When you give people truly useful information, you have shown that you know how they will use that information; otherwise, the information would not be useful. People read information in order to discover what they need to do. Whether your intention is to persuade or to inform people, your writing must clearly show them what actions are suggested or required.

RESEARCH SHOWS...

Readers often understand technical writing by forming a "concrete story or event" in their mind. They do this with even the most technical of documents, because they need to *use* the information in some way. They need to know "agents and action." So, the readers in one study actually rewrote a report by *restructuring* the information to be more active (Flower et al., 1983, p. 45).

1.3 WRITING AND SPEAKING CAN HELP ME DISCOVER WHAT I REALLY THINK

In the process of producing solutions, you will use many different kinds of oral and written communication. As you move through your career, no matter where you work—whether a large Fortune 500 firm, a small consulting firm, or a public agency—writing and speaking can help you plan, develop, and revise your engineering solutions. Writing can solve problems for you as well as for your reader.

Imagine, for a moment, that you must do some research for your boss on nanocrystalline coatings that protect microchips. There are many of these new coatings on the market, and your company needs information on similarities and differences among coatings before investing research-and-development time and money in adapting a coating for use on its chips. What do you do? How do you start? You probably start by reading as much as you can, collecting information on coatings and trying to organize it. Then you have to look for similarities and differences and make judgments about which coatings fit better with your company as products. How do you capture what you are learning? You probably take notes and make summaries of informative articles and websites. So you are already writing, even before you have started drafting the report you have to produce. You are also probably talking to people, asking questions and getting their feedback on your findings.

When you do start drafting the report, you probably begin with the lists you have been making of each coating's qualities (strengths and limitations). If you do not have enough information about a particular coating, you will notice, because your list will not be complete. The simple act of putting down words shows you what you need to find out more about. And as you start drafting and making an argument for one kind of coating over the other, you may notice that you never say why, for example, coating 3 will not work. Your argument is thus not complete, because you had assumed early on in your investigation that coating 3 would never work, but forgot to explain why. And then when you start to explain why, you realize that you were mistaken about one of the attributes of coating 3, and it might work after all.

In these and many other ways, writing helps you see your own thoughts in front of you, so you can clarify and finish them. Without writing, you probably cannot know what you understand and what you do not. Just remember that writing happens over time, as a recursive process of outlining, drafting, and revising. You have to give it time. Computers can help us create and revise documents faster, but they probably cannot help us think faster. And your writing is simply the clear arrangement of your thoughts—after you have rearranged them a few dozen times. Remember that when you see someone else's writing, you are seeing a final product; you cannot see the

KEY IDEA: Writing drafts helps you see your own thoughts in front of you, so you can clarify and complete them.

hours of drafting and thinking that went into that written product. So, do not think that most people write faster than you. Almost no one writes very quickly!

EXPERTS SAY...

According to engineer Henry Petroski, there is a clear connection between the practice of engineering and the practice of writing, since both involve creating something new: "Writing no less than engineering consists of ideas to be realized" (p. 10). In his book *Beyond Engineering* (1986), Petroski draws a parallel between an engineer as a bridge builder and as a writer when he describes an engineer as "the designer of a bridge of words" (p. 10). He also explains how writing about bridge building helps the engineer understand the actual process of bridge building better, because writing is "the test of an engineer's understanding of his own theoretical work" (p. 11).

1.4 MY READER OR LISTENER (MY AUDIENCE) IS ALWAYS MY CLIENT

As an engineering professional, to whom will you be writing and speaking? Those of you who have worked in engineering-related firms, private or public, know that, as an entry-level engineer, you make reports (orally and in writing) to your boss, to colleagues, and sometimes to middle managers from other divisions or offices. For the first few years of your career, you probably will not be writing directly to outside clients. In fact, one survey found that, three years after graduating with a B.S. degree, most engineers were writing more often to people inside their organization than to people outside it (Anderson, 1986). So, in light of this finding, why should you consider your reader a client? Because the other finding from this study was that most of those in-house readers knew less than the writer about the subject matter. So, if you are the writer, your job is to help your readers understand the subject matter and act on that information.

Treat your reader or listener like a valued client—someone with whom you want to maintain a positive professional relationship. And think about how to talk to that client, in writing or by speaking. For example, suppose that you have to present that design for a bus shelter to the city council. How do you explain a design to people who may not look at designs very often? City councils are usually composed of ordinary citizens from many walks of life who happen to care about their city. Do you show them a plan view of your design? A schematic? Will your readers know how to "read" these visuals? Should you refer to the "postmodern use of space"? Will your readers understand that terminology? You would want to examine your communication options ahead of time, including vocabulary, visuals, and even tone of voice.

RESEARCH SHOWS...

In the workplace, chances are good that you know more than your readers—even in-house readers—about the subject matter of your writing (Anderson, 1986). This means that you must explain technical processes, concepts, and vocabulary more than you might think.

1.5 IT IS MY ETHICAL RESPONSIBILITY TO TURN DATA INTO INFORMATION MY READERS CAN USE

Engineers deal with lots of data; they collect and generate test results, experimental results, specifications, standards, all sorts of numbers, and raw facts. In order to make your communication useful for all your readers and listeners, you need to help them understand the meaning and the relationships of your data. Data are just raw facts. For example, consider the statement, "It is 100°F outside." That piece of data will not be very informative to a child who does not understand what "degrees" means or to a visitor from Indonesia who is not familiar with the Fahrenheit system. Data are bits of information (test results, raw facts) that must be interpreted, organized, and synthesized in order to make sense and to have meaning. As one cognitive scientist put it, "Data plus meaning is information" (Reeves, 1996). Your job is to turn the data you produce into useful information by anticipating the needs of your reader. Sure, many of your readers will be technical professionals like you, but they still need to know the relationships among the data you present.

KEY IDEA: Data are raw facts/numbers; information is data made useful for someone.

The information, for instance, that air density changes with temperature can be demonstrated by presenting the correlation between air density and temperature at successive measurements. You would probably use an x–y scatter graph, with a line drawn through the series of points plotted, to convey this information, as in Figure 1.2. If you chose, on the other hand, to convey the information in words alone and presented the temperature readings in one paragraph and the corresponding density readings in another paragraph, you would make it way too hard for the reader to perceive how rapidly air density changes with temperature (especially by contrast with water density, for instance). And you would be very far from imparting knowledge to your reader that he or she could use later on. The x–y graph provides the best context (a visual context) for the reader to understand the correlations. And then your words can enhance and support the visual context: "At standard atmospheric pressure, the density of air drops steadily with increasing temperature." Graphical and verbal presentation of data, working side by side—bingo, you have created information your reader can use.

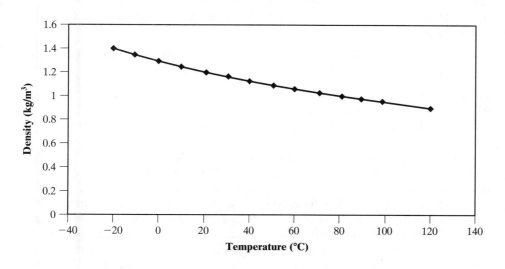

Figure 1.2
Relationship between density and temperature of air at standard atmospheric pressure. *Source of data:* Crowe, C.T., et al. 2001. *Engineering Fluid Mechanics*, 7th ed.

At this point, you might ask, "Why is communicating clearly an *ethical* responsibility?" In fact, the history of technology development is littered with examples of failures brought about, at least partly, by inadequate communication among engineers, management, and others. Edward Tufte and other researchers have written extensively about the communication problems between Morton Thiokol, the rocket manufacturer, and NASA in the hours preceding the decision to launch the space shuttle Challenger on a cold January morning in 1986. The Challenger blew up 73 seconds after takeoff, killing seven astronauts aboard (including the first woman in space, Christa McAuliffe). The analysis Tufte performed on the charts sent to NASA by the engineers the day before the launch is quite convincing: even though the engineers originally warned against the launch, because of possible trouble with the O-ring seals in cold weather, the charts failed to convince NASA not to launch. Even Morton Thiokol management later reversed their conclusion not to launch. The cause of the eventual catastrophe was actually debated by all those technical professionals on that fateful evening, but they failed to make the correct decision. Why? The charts prepared by the engineers failed to convey the danger clearly; they were fragmented, vague about the cause of previous problems with the O-rings, and unconvincing about the link between O-ring failures and temperature. In other words, the engineers did not turn their data into sufficiently useful information, information that would make the right decision irrefutable.

Of course, the bad decisions of January 27–28, 1986, were not entirely due to communication failures; the expense and waste of aborting a launch at this very late date no doubt weighed heavily on the minds of NASA management, and eventually even the MK engineers were persuaded to think more like business-minded managers than like engineers primarily concerned with safety. But decisions can only be made on available information, and the more clearly that information is presented, the better the chance of making a good and safe decision.

Why is it your job to create information that people can use? Isn't it your job only to design or calculate the technical solution to the problem? As a matter of fact, your professional responsibility extends to the use that people make of your solution. Most engineering codes of ethics begin with a statement about an engineer's responsibility to the people her work will presumably benefit. As an example, here is the first canon of the Code of Ethics of the American Society of Mechanical Engineers:

> *Engineers shall hold paramount the safety, health and welfare of the public in the performance of their professional duties.*

For every decision you make as an engineer, every design, process or product to which you contribute, you will need to keep in mind your obligation to protect "the safety, health and welfare of the public." That obligation must be "paramount" in all you do.

If you shirk your professional responsibility to communicate data in language and visuals that your audience can understand, then you are failing to live up to this code. Much of the investigation in the aftermath of the Columbia (2003) and Challenger (1986) space disasters focused on failures to commuicate technical findings appropriately to decision-makers. In both cases, the wrong decision was made; the Columbia attempted to re-enter the earth's atmosphere without its damaged tiles having been replaced, and the Challenger was launched in cold weather and exploded a few seconds later. You can learn more about the role of inadequate communication in these disasters by reading the Tufte selections listed in the Further Reading section at the end of this chapter.

1.6 MY AUDIENCES WILL CHANGE, AND SO MUST MY WRITING AND SPEAKING

One of the challenges of learning how to write for engineering work is that, in school, you mostly write to show what you have learned. Your professors and teaching assistants are not your boss; they are not paying you to solve problems. Neither are they typical readers of your engineering documents. Unlike your future supervisors, clients, and colleagues, your instructors want to know how much you have learned about what they already know. So it is sometimes difficult for instructors to respond to your writing the way typical readers might, with thoughts such as these:

"What does she mean by 'picocuries'?"
"I thought he was going to give me credit for my part of the design."
"This report looks too long to read."
"Doesn't she know we don't use that pollutant-removal process here anymore?"

The reality is that engineers must communicate daily with listeners and readers who have a variety of educational and professional backgrounds. To give you an example, assume for the moment that you are an environmental engineer. For a final report on your study of methods to clean up a particular hazardous-waste site, you may be addressing regulators who are familiar with some of the terminology in your discipline ("*in situ* remediation," for instance), but who do not have a strong background in the science of sampling or of assessing risk. If you then write instructions for the technicians who will implement the cleanup, you will have yet a different audience, one who may not understand the regulatory terms or the science, but who will need clearly written procedures for handling hazardous materials. And sometime later, you may need to help write a description of the remediation process for local news reporters who know very little about engineering. See examples of these three styles of writing in Figures 1.3 through 1.5. Each of these audiences has

KEY IDEA: Your readers (and listeners) will range from people with very technical backgrounds to those with no technical background at all.

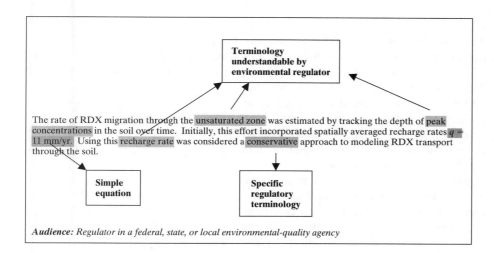

Figure 1.3
Excerpt from a *technical report* on RDX contamination at a site.

Figure 1.4
Excerpt from the introduction to a *set of instructions* on implementing a bioremediation system to clean up explosive RDX in contaminated soil.

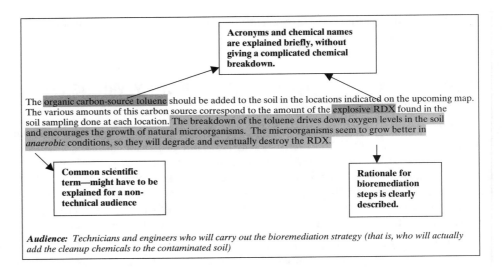

Figure 1.5
Excerpt from a *newspaper article* on cleaning up RDX contamination.

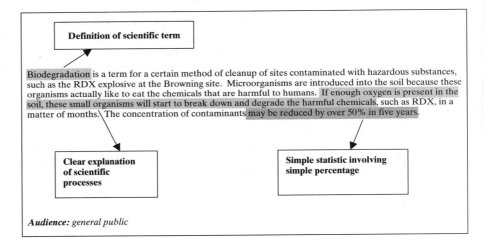

different needs for your documents, needs that must be met if your documents (and presentations) are to communicate meaningfully and get the particular job done.

How many different audiences will you have in your career as an engineer? The answer may depend partly on whether you go into management or not. Chances are that you will become a manager within five years of graduation from college. At the very least, you will be writing for and speaking to people whose role in your organization falls into one of the three categories defined by Paul Anderson (1986): Decision-makers, advisors, implementers.

Decision makers make things happen in a company or organization. They are ordinarily top management or those managers with some budgetary control. They

Decision Makers

- Spend most time managing resources and people.

- Are almost always in communication with someone.

Need summaries and recommendations.

are "big-picture" people who must allocate resources, keep an eye on the bottom line, and keep the organization moving forward at a healthy, (but not out-of-control) pace. They want summaries of information and quick answers to questions such as, "How much will this solution cost?" and "How long will the solution take to implement?" They want to read conclusions and recommendations, not report sections containing lots of details. They want to listen to talks that get to the bottom line quickly.

Advisors are indispensable to decision makers, because advisors *do* care about details and methodology. Their job is to advise the decision makers about the credibility of solutions and of the procedures used to develop those solutions. They want to see descriptions of procedures and discussions of results. They will even check the appendices of reports to make sure that the investigation or proposal follows good science and that the computations are correct.

> **Advisors**
>
> - Check details.
> - Use their subject-matter expertise to help decision makers make decisions.
>
> *Need methodology and procedures descriptions.*

The final audience type, implementers, usually do not have much decision-making power. Instead, they have to do things on the basis of your writing: recalibrate a machine, build a new component to filter out "bad" air from a manufacturing building, etc. They are often technicians. These readers need clear instructions and rationale for the changes they will be making. They do not need theoretical details, finan-

> **Implementers**
>
> - Do hands-on work.
> - Are often out in the field, in the lab, or on the factory floor.
>
> *Need clear instructions and explanations.*

cial justifications, or lengthy discussions of procedures used to arrive at the changes.

There are other audiences inside your own organization who may not quite fit into those categories: colleagues (engineers at the same level as you), support staff, and sales and marketing staff. The first audience is easy to communicate with; the others need clear explanations of concepts and terminology with which they are probably not familiar. And then there are the audiences outside your

---

Practice Question: Differing Audiences

As an audience for your writing or speaking, how does a workplace engineer (manager or nonmanager) differ from an academic engineer (such as your engineering professors or instructors)?

Suggested Answer

Workplace engineers are extremely busy, typically involved in a number of projects, and usually responsible for keeping costs as low as possible for the company and the client. Managers often do not do any engineering work, but rather manage people and other resources. Academic engineers are usually most interested in underlying causes of problems that need solving, and they want to pass their expertise on to younger engineers. Your instructors want to know how much you know and how well you know it; workplace engineers want to know what you know that can help them solve a problem.

organization: customers, regulatory agencies, financial institutions, vendors, and perhaps the news media. As a manager, you will be called on to communicate with at least the first four of these audiences and possibly with the last one as well.

SUMMARY

- For engineers, writing is not an extra set of skills that it would be nice to possess, but rather an integral part of doing engineering work every day.
- Writing helps you solve problems by helping you discover what you really think about the connections among the data and facts you have collected.
- *Data* are raw facts and numbers; *information* shows relationships among data and thereby becomes useful for solving problems. Your professional responsibility as an engineer is to turn data into information people can use.
- You will have many different audiences for your technical writing and speaking, and you must learn to adjust your language and graphics to fit the understanding of the particular audience.

PROBLEMS

1.1. Interview an engineer in the workplace. Choose any sort of engineer; he or she may or may not be working in an engineering firm. Ask him or her to describe a typical day in their workweek: what tasks does he or she perform and how long does each take? Ask especially about how much time your interviewee spends writing and reading, including e-mail and other forms of digital communication. Write up a short summary of the interview and compare with summaries written by other students. How much time (expressed as a range) do engineers spend doing communication tasks?

1.2. Talk to a family member or a friend who is not an engineer and describe a project you hope to work on, either at work or in college. Or describe the latest concept you just learned in your engineering studies. Be as specific as possible. Then, answer the following questions:

Does your listener ask questions as you describe the engineering project? If so, write down those questions.

Can you characterize the questions? Are they primarily requests for definitions, for more description?

1.3. Based on the questions asked by the non-technical person you interviewed, write a short analysis of the types of information needed by a non-technical audience.

1.4. Make a sketch of a particular piece of lab equipment in an engineering laboratory. If you don't have access to a lab, choose another available piece of hardware equipment (such as an electric drill or a coffee maker). Show the sketch to a non-technical person and ask him or her to describe how the piece of equipment works, based on your sketch.

1.5. If the non-technical person had trouble describing how the piece of lab or other equipment works, write a brief analysis of why. How might you redo the sketch, knowing what you know now? If the person had no trouble, write a brief analysis of the strengths of your sketch in communicating practical facts about the piece of equipment.

1.6. Chances are, your engineering instructor is which kind of audience for you?

Decision maker

Advisor

Impenter

1.7. Does the following sentence contain "data" or "information"?
Subjects in the behavioral experiment had a 90% likelihood of pushing the blue rather than the red buzzer.

1.8. Does the following sentence contain "data" or "information"?
The findings from this experiment reveal that the number of total fish in the Hudson River has decreased by 23% since the 1950s.

FURTHER READING

Anderson, P.V. 1986. "What Survey Research Tells Us about Writing at Work," in *Writing in Nonacademic Settings*, Odell, L., and Goswami, D., eds. Guilford Press: New York.

This book is a groundbreaking work that looked seriously for the first time at writing in places other than schools. Paul Anderson pioneered research that looked at writing from the point of view of the reader rather than the writer. His essay in this volume is a clear and readable presentation of his study of alumni of Miami University (Ohio) and their experiences with writing at work.

American Society of Mechanical Engineers. November 2006. *Code of Ethics of Engineers.*
http://www.asme.org/NewsPublicPolicy/Ethics/Ethics_Center.cfm

You may see an Interpretation of the Code of Ethics on the ASME site:
http://files.asme.org/asmeorg/NewsPublicPolicy/Ethics/10938.doc

Crowe, C.T., Elger, D.F., and Robertson, J.A. 2001. *Engineering Fluid Mechanics*, 7th ed. John Wiley & Sons: New York.

This is the textbook from which I took the data to construct the graph in Figure 1.2. You can learn a lot about good technical writing by looking at a good textbook for undergraduate students (who have, after all, a variety of educational backgrounds). Notice the way information is broken up into many sections with headings and visual items such as shaded boxes and illustrative figures.

Flower, L, Hayes, J.R., and Swarts, H. 1983. "Revising Functional Documents: The Scenario Principle," ... in Anderson, P.V.; Brockmann, J.R.; Miller, C.R., eds, *New Essays in Technical and Scientific Communication: Research, Theory, Practice.* Farmingdale, N.Y.: Baywood Pub. Co.

Everyone knows that scenarios are good to construct when you are making a video or film, but most technical writers still balk at the idea of telling a story. This article demonstrates persuasively that even knowledgeable readers will reconstruct a passage in their mind to resemble a story with "agents and actions" (p. 45). So we may as well write that way, at least in functional documents such as instructions and procedures.

Petroski, Henry, 1986. *Beyond Engineering: Essays and Other Attempts to Figure without Equations*. St. Martin's Press: New York.

This small book is much beloved by both engineers and communication professionals who are interested in convincing other people that engineering and writing are not mutually exclusive activities; they actually support each other. Written by an engineer with much experience in both industry and academia, the text is down to earth and conversational.

Reeves, W.W. 1996. *Cognition and Complexity*. The Scarecrow Press: Lanham, MD, and London.

This volume is a wonderful little book that explains very readably what cognitive science has to say about how people process information and how we are all coping with the anxiety of information overload. Chapter 2 explains succinctly the similarities and differences among "data," "information," and "knowledge" and why it is important to use clear presentation to turn data into knowledge.

Spilka, R., ed. 1993. *Writing in the Workplace: New Research Perspectives*. Southern Illinois University Press: Carbondale, IL.

This collection of articles extends the work of *Writing in Nonacademic Settings* to present more conclusions about the interplay between writing at work and working. In particular, Spilka's article ("Moving between Oral and Writing Discourse to Fulfill Rhetorical and Social Goals") clearly shows how both speaking and writing help individuals and organizations achieve their professional goals.

Tufte, E.R. 1997. *Visual Explanations: Images and Quantities, Evidence and Narrative*. Graphics Press: Cheshire, CT.

In a chapter entitled "Visual and Statistical Thinking: Displays of Evidence for Making Decisions," Tufte presents a detailed analysis of the failures in communication that helped lead to the 1986 Challenger disaster. All the charts and memos failed to make verbally and graphically clear the connection between cause (cold weather) and effect (brittle O-ring seals).

Zinnser, W.K. 1988. *Writing to Learn*. Harper & Row: New York.

This wonderful, easy-to-read book presents the connection between writing and thinking. Especially useful to engineering students are Chapters 2 ("Writing across the Curriculum"), 9 ("Writing Mathematics"), and 10 ("Writing Physics and Chemistry").

Discovering Ideas and Facts: Researching

2.1 INTRODUCTION

Whatever the subject of your writing project, you will begin by collecting and organizing a lot of information. You need two kinds of information, both of which often overlap:

1. Information to help you understand the issues, facts, problems, and solutions:
2. Information to be presented to your audience.

Most of the information of the first type is derived from your research. Information of the second type may include what you already know as well as some of what your research reveals. The second type of information is what ends up in your document or presentation, and your job is to figure out how much of your research needs to be passed on to the reader or listener.

You cannot do productive research unless you know the purpose of the research project and the audience who needs to know about it. Let us say you are doing a report on noise-reduction methods for an existing apartment complex whose tenants have been complaining that their apartments are noisy. No matter to whom the report will be directed, you have to gather information about acoustic sealants, absorptive materials, how sound waves bounce and reverberate, etc. But if the report is for the apartment-complex owner, rather than the testing lab, you may need to gather somewhat different evidence. For example, you might want to incorporate in your report the graphics from a reputable website showing how sound waves act when they hit various surfaces. (You will, of course, credit your sources fully in the report.) The testing lab, on the other hand, would need no such graphics.

College instructors are usually experts in the subject you are writing about for their classes, so it can be difficult to know how much background explanation to include in assignments for them. On the other hand, instructors also want to know how well you understand the concepts and how much work you have done, so they may actually want more explanation and illustration than a workplace audience would. Discuss the issue of appropriate audience with your instructor; be sure to ask him or her if you should have a workplace audience in mind, such as an owner, a government agency, or a consulting client.

2.2 FRAMING YOUR WRITING PROJECT

This section discusses some typical writing assignments in engineering classes, along with the procedures for accomplishing the assignments. Note that each step involves reading, writing, or speaking, but only the italicized steps indicate what you may regard as "real" writing. Each of the italicized steps will generate at least one section of a report or proposal.

2.2.1 Lab Reports

Most lab reports document what you did in the laboratory as you followed a particular experimental or testing procedure. You will need to record carefully all data you collect and to note at the time any deviations from given procedure. Your lab report will really benefit from careful and timely note taking. Keep a notebook with you at all times in the lab. Be sure to record all that you observe, even if your teammates are also supposedly recording.

Here is a set of steps for writing lab reports. As with all writing procedures, you may have to loop back and repeat some of the steps (e.g., you may want to graph your data before thinking about what they show). The steps are as follows:

1. Define the law or hypothesis you are trying to prove.
2. Record the data as you go through the procedure.
3. Organize the data, and think about what they show.
4. Create figures and tables to display the data.
5. *Write a first draft of the Results section.*
6. Check your findings with your teammates (if applicable).
7. *Revise and rewrite the report.*

These steps may not be entirely linear, but you definitely want to accomplish steps 3 and 4 before attempting step 5. Do not try to start writing the Results section until you have created the tables and figures that display your data in meaningful relationships. Look at Figure 2.1. In this case, once you have plotted the values for engineering stress and strain and for true stress and strain, you can easily see at what point the values start to differ. Then your verbal description of what happened becomes easy to write: "At low levels of applied load, the values of engineering and true stress and strain were nearly the same. The differences increased, however, as the magnitude of the applied load increased."

Here is an outline of the sections of a typical lab report, whether academic or industrial:

Introduction (sometimes includes a Background section)	**Appendices**
	Measured Data
Test Procedure	**Sample Calculations**
Results and Discussion	**Notation**
Conclusions	

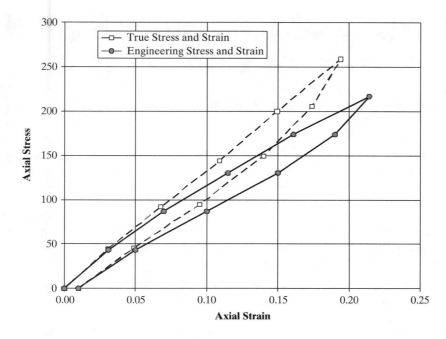

Figure 2.1
Measured stress–strain curves for rubber specimen.

Some formats for lab reports divide the Results and Discussion section into two sections, and sometimes a Recommendations section is added at the end of the report. But generally, reports on laboratory tests follow an outline close to this one.

In academic lab reports, the reader (your instructor, usually) is most interested in whether you performed the test correctly and learned something from it. Thus, any deviations from established procedure should be carefully noted, and the discussion of the results should show your attempt to organize, see in relationship, and then understand the *significance* of the data you collected. In industry lab reports, the reader is often someone outside your department or organization who needs the information in a hurry in order to make decisions. Thus, for example, the director of engineering for a tire-manufacturing company may need the results of the rubber test in order to assess the strength of the new tire designs being produced by the engineers.

Be sure to think about which engineering problems your findings are likely to help solve; mention those applications in the Conclusions section. If things did not go as planned in the lab, do not worry. Writing up the problems with a particular experiment or test can be very useful to others interested in the same experiment. Failure in engineering can teach us a great deal, and the ideal place for failure is in the lab or under simulated conditions.

2.2.2 Research and Recommendation Reports for Decision Makers

Apparently, there is often a big difference between how students and instructors conceive of a research project. Students feel they are simply supposed to show that they know something, whereas academicians use and do research in order to solve a problem or test a theory (Ford, 1995). Those latter goals require the ability to analyze and synthesize information, not just collect and regurgitate it. And, especially in engineering research, you need to exercise judgment as well as analytical skills in order to develop the best and the safest solution.

Students think that in their research papers, they should show they know something. College and university instructors, on the other hand, believe that the purpose of research is to test a theory or explore a problem (Ford, 1995). This belief means that instructors want to see evidence of synthesis, analysis, and judgment. Clarifying your method of investigation shows that you are *thinking*.

If your research topic is given by your instructor, then you have a head start on the first step in each of the procedures listed next. If not, you will have to come up with a suitable topic for investigation. Presumably, your instructor wants you to demonstrate good research skills as applied to real engineering problems. For sources of good topics to investigate and write about, think about the engineering problems you see all around you—problems with infrastructure, mechanical devices, cars, electronic devices, airplanes, etc. Here is a list of possible topics for a recommendation or research report:

1. Choosing sustainable building methods and materials for a home owner.
2. Evaluating wastewater-treatment methods for a pig farm.
3. Investigating landing-gear systems for a new-model business jet.
4. Evaluating launch vehicles for a communications company's satellite.
5. Comparing robotic-drive platforms over variable terrain for construction automation.
6. Examining the feasibility of tissue engineering for artificial organs.

Notice that in all these topic designations there is an implied client—for example, the owner of the home in #1 or the communications company in #4. In the last two topics, the implied client is the research-and-development division of a company (a construction-automation firm or a biomedical firm, respectively). Engineering work is usually done to solve a particular problem for a particular decision maker or set of stakeholders. Unless you go into basic research, you will generally not be doing research without a specific audience for the results.

The upcoming box shows a procedure for researching and preparing reports for decision makers and stakeholders. The two writing stages (drafting and revising) are not numbered because you can work on them *at any point in this process.*

	Recommendation Report	**Analysis Report**	
	1. Define the problem or need.	1. Define the problem or need.	
	2. Identify the research question.	2. Identify the research question.	
	3. Define the research methodology.	3. Define the research methodology.	
Write a first draft.	4. Establish selection criteria.	4. Synthesize and interpret information.	*Write a first draft.*
	5. Synthesize and interpret information.	5. Create figures and tables to display the data.	
	6. Create figures and tables to display the data.	6. Reach conclusions.	
Revise and rewrite the report.	7. Reach conclusions.		*Revise and rewrite the report.*
	8. Make recommendations.		

Adapted from Lay, M., et al. 2000. *Technical Communication.*

All reports present analysis of a problem, but a recommendation report is specifically written to help decision makers take action. Most often, the action will be a means of solving a problem, and you (the researcher) will typically be investigating and choosing between courses of action. For an analysis report, the procedure for investigation is the same except that the end result will probably not be an explicit recommendation. In this case, the investigator is probably not choosing between courses of action, so selection criteria are usually not necessary. You might do a report analyzing the various causes and consequences of a particular engineering disaster, for instance, without taking the further step of evaluating steps to prevent such a disaster from happening again. If you do not take that final step, then you would be informing many different stakeholders (government officials, citizens, engineering standards associations) about the disaster's causes, so that further work might be done to develop preventative measures later. If you were writing a report evaluating those preventative measures, then you would be writing a recommendation report, perhaps for a government policy-maker.

In order to recommend a solution or to successfully analyze a problem, you will have to be very precise in your definition of the problem. For a report that analyzes several possible foundations for a planned microchip facility, you might define the potential *problem* for the facility as finding a foundation that does not experience excessive settlement. Too much settlement of the building can damage the delicate machines that produce microchips. You would then need to define "excessive"; as a professional engineer, you would work with the facility's owner to develop the allowable limit. If you defined "excessive" as "settlement greater than 0.25 inches," your project would then investigate only those foundations that reach deep enough into the soil to incur settlement less than this amount.

The *research question* thus becomes, "Which types of foundations are deep enough to cause only that minimum amount of settlement (<0.25 inches) in the clayey soils around the microchip facility?" Answering this question leads you to choose three types of foundation to evaluate, all types of *driven piles* (piles driven deep into the ground): concrete piles, hollow steel piles, and concrete-filled steel piles. So, now you have a refined topic that you can begin researching in earnest. The trick with research reports is to remember that you have to do some up-front research simply to determine your precise topic. Students can quickly get into trouble if their topic is too broad; you cannot, for instance, investigate all types of foundations for all types of buildings. And in the real world, you would hardly ever need to!

How would you collect information about driven piles? Well, you would probably consult an engineering reference book and surf the Web for information on piles. You might also check some engineering journals to see whether any tests have been conducted on the various types of piles. And you would try to talk with engineers who have worked with these piles on other projects. It would be ideal if you could conduct your own tests in a laboratory. You would then be doing what is called "original research." But chances are that you will find the information you need in published sources or by interviewing foundation engineers (experts on the subject matter). You will want to determine early in your project just which research strategies and sources are best for your project and possible for you to use. Here is a list of common research methods used by engineers in academia and in industry:

1. **Conduct tests.**
2. **Observe.**
3. **Solicit expert opinion.**
4. **Collect and synthesize information from published and unpublished sources.**
5. **Make calculations.**
6. **Create preliminary designs.**

Every researcher, whether at a university or in industry, whether a graduate or an undergraduate student, typically uses at least method #4 and often uses method #5. But the other methods may also be possible and useful for you. If, for instance, you are doing a project on possible solutions to the traffic problems at a particular intersection of roads, you do not need to use automatic counting devices to count the cars going through that intersection at peak congestion times, such as rush hours. If you manually count the number of cars within a half-hour interval, for instance, you will have some solid and specific data about the extent of the congestion problem.

Looking again at the procedure involved in producing a recommendation report, you may wonder why the word "scope" does not appear in any of the listed steps. Much engineering work is defined in terms of scope; the Scope of Work is often a section of a proposal. Actually, accomplishing the first three steps—defining the problem or need, identifying the research question, and defining the research methodology—determines the scope of your investigation, the particular piece of the whole problem for which you will try to find a solution. So, defining your scope of work actually involves at least these three critical steps. You should think specifically about where and how you will get information *before* you finalize your scope of work.

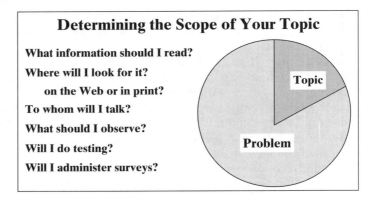

Let us return to the project of investigating alternative foundations for the microchip facility. Because the goal of this project is to choose one type of foundation, you must first understand the constraints put on the possible solutions and then develop a list of *criteria* by which to compare alternative solutions. As stated previously, a big constraint for the foundation investigation is minimal settlement: an allowable limit established by the facility owner's engineers. Since several types of piles can meet that constraint, other qualities come into play in choosing the best foundation—structural strength and cost, for instance. These criteria are what help you select the best alternative for this application.

KEY IDEA

Criteria such as these measure the feasibility of solutions:

- time to implement
- cost
- size, weight, and other physical aspects
- adaptability and compatibility
- efficiency (as a ratio)

How do you define appropriate evaluation criteria?

- **Define your criteria by breaking down the categories into specific solution goals for your project:**
  - **What is the ceiling for cost?**
  - **How big is too big?**
  - **What does "efficiency" mean in this case?**
  - **How long is an acceptable implementation time?**

The rest of the report-developing steps involve doing the work you have set out for yourself. Table 2.1 shows how each of these procedural steps helps you develop particular sections of a standard recommendation report. You can begin writing your first draft at any point during the first six information-gathering and interpreting steps.

Here is how the parts of a report evolve from your writing and researching procedure: An explanation of how you defined the problem and identified the research question will probably appear in your *Introduction*. An explanation of the criteria by which you evaluated possible solutions may have its own section, *Selection Criteria*, as might the outline of your research methods in a *Methodology* section. In short reports, the last two sections might be folded into the introduction, but it is wise to limit the introduction to three pages. Your display and analysis of the data and information you have collected will constitute the body of your report in

Table 2.1 Mapping Particular Cognitive and Writing Activities to Sections of a Typical Report

	Cognitive/Writing Activity	**Section of Report**
	1. Define the problem or need	**Introduction**
	2. Identify the research question	**Introduction, Methodology**
	3. Define research methodology	**Introduction, Methodology;**
Write first draft	4. Establish selection criteria	**Introduction, Selection Criteria;**
	5. Synthesize and interpret information	**Description of Alternative Solutions, Comparison Of Alternatives**
	6. Create figures and tables to display data	**Description of Alternative Solutions, Comparison Of Alternatives**
	7. Reach conclusions	**Conclusions**
Revise and rewrite	8. Determine recommendations	**Recommendations**

several sections that capture the essence of your method—*Description of Alternatives* and *Comparison of Alternatives*, for instance. In scientific papers, this analysis is usually presented in a section called *Results and Discussion. Conclusions* make up their own section. Sometimes recommendations are included as well—in such cases, the section is called *Conclusions and Recommendations*—and sometimes you will present them in their own section, *Recommendations*.

There is one more, very important, section that usually begins the report: *Executive Summary*. In fact, the executive summary is the most important section of the document; as its title indicates, it may be the *only* section that a busy decision maker reads. Thus, the executive summary should be a stand-alone document; if it includes graphics, for instance, they should not be numbered in sequence with the graphics in the rest of the report. Neither should the executive summary refer specifically to any graphic or page number in the rest of the report. To offer the busy reader the bottom line, the executive summary contains a little bit of all the types of information covered in the report; it presents the problem, the alternatives investigated, and the results and conclusions, but *in far less detail*.

The trick in writing an executive summary is to provide good transitions between sentences, since you are hopping quickly from one topic and type of information to another. Here is the final paragraph from a page-long executive summary for a report comparing deep-foundation types for a proposed microchip manufacturing facility. The boldface words indicate transitions that guide the reader swiftly through this recommendations paragraph:

> Advantage Consultants recommends that EMM, Inc., use hollow steel-pipe piles that are 55 feet long and 14 inches in diameter for their foundation. The final foundation layout costs $405,000, making it more expensive than some of the other solutions. **However**, this solution allows the least amount of settlement while staying within the budget. **Based on the presented research**, Advantage Consultants is confident that EMM, Inc., will achieve the best building foundation with hollow steel-pipe piles.

The executive summary comes first in a document, but it is usually written last, after the rest of the report has been drafted.

For an analysis report, you would probably have fewer sections—no selection-criteria or recommendations sections, for example. For a report on drinking-water treatment methods, you would probably have a *Description of Methods* section. You might also have a *Comparison* section in which you listed the pros and cons of each method, without regard to strict technical, regulatory, or financial criteria.

Understanding how your research and analysis will be written up should encourage you to start drafting before you have quite finished the research. For example, you can write a *draft* of the introduction before you have finished collecting information, because you already know the problem, your goal for a solution, and the approach you are taking to find that solution. You even already know the criteria you will use to evaluate alternative solutions. Remember that this draft will need to be revised later, but drafting the introduction early helps you envision and plot the course of your investigation. Meanwhile, you have written part of your report before you have even collected all the data!

Practice Question: Defining Topic and Scope

At work, you are usually given the topic and the scope of work for a project, but when you write an unsolicited proposal or when you are in school, you often must define the topic and scope for yourself. How do you go about defining them?

Suggested Answer

You begin by collecting as much information as possible about the general topic; then you try to narrow the topic by doing more focused research, often by using search engines. You want to find out whether you can narrow the scope enough to be able to do a thorough job of researching and synthesizing information within a specified amount of time. Before you decide on a particular topic to write about, you should already have done some research.

2.3 RESEARCHING

You probably think of research as the process of tracking down published information. Actually, you can also uncover much relevant information through unpublished sources, especially through interviews with subject-matter experts (engineers and others who know more than you do about a particular topic or situation). And you can also do your own original research, even if you are an undergraduate student.

2.3.1 Published Information

Fortunately, tracking down published information is relatively easy and fast on the Web. Not all the websites you uncover, however, will have reliable, credible information. You always need to evaluate these sites by noticing the domain (*.com, .gov*, etc.), as well as any other publication information you can gather, such as author, date, and sponsoring organization. Basically, you want to discover who published the site and whether the individual or the organization has any bias or any proof of credibility as an expert. You also want to find out how recently the site was updated.

Of course, .edu and .gov sites may also have questionable sites (though more rarely). Some of these educational and policy organizations are more questionable than others; some accept money from special interest groups, and therefore their information is somewhat tainted. Be sure to check out the sources that any site references—where does the trail lead? Google comes in very handy for searching opposing points of view or credible authors/organizations who take exception to the seemingly credible information you have found. The point is to do balanced research, checking all sources as thoroughly as possible and using the strategies in the checklist at the top of the next page.

Searches, whether via an electronic database or a Web-based search engine, are an art form in themselves. You can waste a lot of time by using keywords that are too broad. For example, if you type in "pollution," you will get hundreds of thousands of hits. Adding another word like "air" would help somewhat, but you would still have an unmanageable number of sources to check. Your topic should not be that broad anyway. If you are interested in air pollution, you could do a research paper on the contribution of "sewer gases," let us say, to the problem. And you might have to add even more keywords to make that search productive.

Strategies for Researching

- **Be as specific as possible in your search.**
- **Learn to access research databases and peer-reviewed journals. You cannot find credible research studies only by using Google.**
- **Choose the best search engine:**
  - **Google loads faster than other engines.**
  - **Google does not do relevance rating.**
- **Consult a librarian.**
- **Set a timetable for research.**
- **People are also sources of information, but they have to be reliable and credible.**

As with writing, the process of researching may reveal several items. You may discover what it is you are really interested in (not just air pollution, but radon in particular, let us say, and especially how dangerous it can be in certain amounts). Or you may discover that what you are interested in involves proprietary information that is simply not available. In that case, you would be wise to adjust your topic. Refining your topic early, before you have gone too far with research and writing, is vastly preferable to having to change your topic later.

Here are two of the best sources for engineering information:

Engineering Index (EI Compendex Plus)
This index contains more than 1 million citations, with abstracts from more than 4,500 journals, reports, and conference proceedings covering fields of engineering and technology. It is the best overall source for scholarly engineering information.

Applied Science and Technology Abstracts
Covers 350+ key, international English-language periodicals in the applied sciences and technology. It is a good source for practical (that is, not as research-oriented) information.

2.3.2 Unpublished Information: Interviews

You can learn a lot by talking to people, especially experts in a particular field or experienced managers in any organization. Because experts and managers are usually decision makers and are extremely busy, the trick is to convince them to give you the time. Being a student is an asset, since most professionals want to help students: Do not expect, however, to take more than 30 minutes of their time. Interviewing for information is an art. If you can master this art, you can learn a great deal from those professionals who have themselves taken years to learn what they know.

Showing interest in the interviewee and enthusiasm for the subject are critical ingredients of a good interview. One way to show interest is to lean slightly forward, toward the "interviewee" or at least to sit upright. Another critical factor is coming in prepared with good, focused questions. If your interviewee talks a lot and wanders off the subject, you may have to carefully and politely push the conversation back on

> **Collecting Information: Interviewing**
>
> - Gather necessary information.
> - Prepare effective questions:
>   - Focused
>   - Few yes-or-no questions
>   - One at a time
> - Don't interrupt.
> - Allow time for response.
> - Convey interest:
>   - Tone of voice and body language
> - Write thank you letters.

track. But also be prepared to simply listen when he or she is on a roll; listening well usually helps you think of more good questions to ask. Never record a conversation (on the phone or in person) without asking for permission. If you are taking notes, do not lose your focus or slow things up by writing too much down; wait until after the interview, and then *get the information down on paper* as soon as you can.

RESEARCH SHOWS...

How do you know whether someone is an expert? According to Leahey and Harris (1997), aside from possessing "an extensive knowledge base" in their subject area, experts can quickly distinguish relevant from irrelevant information and focus on the former. Remember this caveat when you interview experts: Cut to the chase with your questions. Experts are also flexible and adaptable, "adjusting their strategies to fit a constantly changing situation" (p. 246). In other words, you can learn a lot from experts just by watching them at work. Finally, experts can "simplify, chunk, and organize information efficiently"—just as *you* should.

2.3.3 Unpublished Information: Surveys

There is also an art to designing good surveys. You want to capture *only* the information you want, and you want people to actually complete the survey. So your design should be simple and attractive, and the survey should be as *short* as possible. Never ask people to write out their thoughts for more than one question, and be prepared not to get anything of substance from free-answer questions. The success of your survey should depend on the simple-to-answer questions, not on any essay-type answers you try to elicit. Always consider getting demographic information such as age, sex, and employment, but usually you should not ask for a name, since you want people to feel free to express their real thoughts without worrying that those thoughts can come back later to haunt them. (Remember how many public officials have bemoaned the fact that the press sometimes takes their remarks out of context.) Write an introduction that states the purpose of the survey and the importance of participation. And always include a written thank-you statement.

> ### Collecting Information: Surveys
>
> - **Use a random, but representative, sample.**
> - **Ask short, mainly yes-or-no, simple questions.**
> - **Ask very few open-ended questions.**
> - **Avoid vague or overly general words.**
> - **Be cautious about using jargon and slang.**
> - **Use an easy-to-complete, attractive design:**
>   **—No more than one page.**
> - **Collect answers yourself whenever possible.**

You can collect some useful information from even a small sample, as long as that sample is random. For instance, for that bus-shelter design project, you might survey riders of the municipal bus system about their preferences in bus shelters, including the features they would like to see. You would probably survey only bus riders (the users of the system), but you would try to vary their age, sex, educational background, and socioeconomic level by administering the survey at several bus stations in very different parts of town.

2.3.4 Evaluating Sources of Information

The integrity of your writing—of the information presented in your writing—depends to a large degree upon the integrity of the sources of that information. If a particular source is an expert whom you have interviewed, you can probably rely on his or her credibility, although you still might want to interview at least one other expert in order to gather corroborating information. If the source is a book or article published by a reputable publisher or a recognized technical journal, and if the work has been favorably cited by other researchers, then you probably have a credible source. But lots of information about technical subjects is available on the Web, and you know that not all the sites you see contain reliable, unbiased information. You must, of course, evaluate these sites by first checking out the domain: *.com* indicates a commercial site; *.org* and *.edu* usually indicate noncommercial sites, and *.gov* indicates a government site. Information on a university or public-agency site will tend to be more reliable than information posted by a private company or a special-interest group. Companies want to sell products or services, and special-interest groups want you to join their cause; on the other hand, universities have a long tradition of encouraging peer review of faculty and student work, and public agencies are accountable to citizens. So, what is published on the latter two types of sites has probably been through an unbiased review process and is therefore usually considered to be reliable information—unless, of course, you are looking on the home page of a faculty member who has not published the information anywhere else. In that case, the site is just as suspect as any site that presents unreviewed opinions. Look for a tilde followed by the person's name (e.g., ∼ *smith*) in the directory path of the URL; this designation is a tip-off that the page may be located in a personal directory.

Trace backward in the URL containing information you are scrutinizing in order to discover the home page; somewhere in the site, there should be information about how and by whom the site was developed. Look at the "About" page and check out the site map. Does the organization make sense? Are there some dead links? Internet robots and spiders are helpful aids in the search for good information,

Evaluating *All* Sources of Information

1. Does the information answer important questions for your research?
2. Is the information current? Can you determine the date of publication?
3. Are presented facts and figures from reliable sources?
4. Can you detect any biases in the way that the information is presented?
5. Can you determine the author's credentials and qualifications?
6. Is the publisher or sponsor credible? For WWW sources, check the domain.

but be careful about the URLs they gather, as they look everywhere, even on sites that are no longer maintained or are under construction.

You must also evaluate websites by uncovering any other publication information you can, such as information about the author, date of publication, sponsoring organization, and purpose of publication. Determining some of this publication information can be tricky. If there is no explicitly provided publication date, look on the document-information screen for a notation of the last date of modification— "Last updated on *mm/dd/yy*." If there is no date anywhere, you must document the date you accessed the site. Otherwise, you will have no proof that that site existed and contained the information you may reference in your writing. So, be sure to keep a dated record of your Web sources, including pathway information, so that you can return to them. You will need this dated record in order to compile a proper list of references and to cite sources responsibly in your documents.

Evaluating *Internet* Sources of Information

1. Is the site updated regularly?
   —Check dates.
2. Is the site well designed?
3. Is the writing grammatical and clear?
4. Does the source make any biases absolutely clear? Does it make its purpose clear?
   —Check the "About" page.

Practice Question: Evaluating Sources of Information

If you were doing a report on the economic and environmental benefits of using acid-gas injection instead of sulfur recovery to dispose of the acid gas produced by natural-gas production, would the following website be relevant to your report?

http://www.epa.gov/air/oaqps/airtrans/acidrain.html

Suggested Answer

The EPA article at the website provides some good background information on acid rain, which is produced by pollutants such as those found in natural-gas reservoirs. So, you might use the article in the background section of your report. If the focus of your project is on the benefits of acid-gas injection, however, you will want to move beyond background information quickly. Find sources with credible information on this process, how it works, and how much it costs. Don't get hung up on the lengthy literature about acid rain.

2.4 MANAGING AND DOCUMENTING INFORMATION

Once you have identified and evaluated your sources of information on a topic, you have to keep track of them and determine whether and how the information you have gathered can be used in your document. Let us assume that you are researching foundation alternatives for a planned microchip facility on the outskirts of your town and have determined that three types of driven piles—concrete piles, hollow steel piles, and concrete-filled steel piles—shall be evaluated as possible solutions. We suggest that you follow these steps, because managing information is a **four-step process**:

1. *Evaluate* **the sources and the information itself.** If you have determined that the websites or articles from which you have obtained information on driven piles are reliable sources, you still have to decide whether you are going to use *all* the information you have collected or only the information on two of the three driven piles you started out to examine. By using an initial list of criteria, you may have already determined that concrete-filled steel piles are too expensive and would have to be trucked to the site from very far away (adding even more to the cost). So you would stop collecting information on concrete-filled piles and use the information already collected to explain in your methodology section why you ended up investigating only two types of piles. You can often determine relevance by checking the information against your outline, but make sure that your outline is keeping up with the changes you are making in your scope of work.

2. **Informally, but thoroughly,** *document all sources* **of information.** Even if your instructor or boss has not specified that you include a list of references, you should create one—if only for your own work later on. Chances are that you will be working on a similar topic in the not-too-distant future, and you want to have a record of where you got the most helpful facts and data. Also, if you use information from that source in what you write, then you MUST cite it or you will be plagiarizing (see Section 2.5). So, whether you format a formal list of references or not, always keep a dated record of what information you found where. Use index cards or a word-processing file to keep track of sources. Collect these facts:

author
date
title of work
(title of larger work)
publication information

For electronic sources, add these facts:

electronic address
date of your access

Ask your instructor or boss which referencing style to use. Even within the field of engineering alone, there are many slightly different styles used. For example, some journals want a list of references in the order in which you cited them, with superscript numbers as textual citations, while others want an alphabetized list, with the textual citations containing only the last name of the author(s) and the year of publication. If you are not given a reference book that specifies citation guidelines, consult the website of your field's major professional society and check for something like "A Guide for Authors." The main point here is that when you are in the middle of research, you want to get these facts down and save them in whatever format is most convenient, but eventually you will have to format the collected information according to the required referencing style. You may as well alphabetize your documentation right from the start, for instance, if that is how you will organize your reference list in your report or paper. Please don't waste brain cells in trying to memorize a documentation system. Keep a reference book handy (see suggested guides in Further Reading).

3. *Summarize in your own words* **the information from each source.** When you immediately write a summary of a piece of writing you have just read, you begin the process of synthesizing the information and extracting out of it what is important for *your* work. Especially when you are less experienced and when you have been reading a lot at one sitting, you quickly lose the ability to distinguish between useful information and unnecessary or bad information. You end up reading blindly, and later you cannot remember whether or why that article seemed useful for your report. Break out of the passive mode; become an active thinker simply by writing down your own outline of the article's contents and your own thoughts about its quality, findings, and relevance. After summarizing an article, you are one giant step closer to writing your own report than you would be if you just tried to memorize the article's facts or to take notes on what it says.

Try also to link the information in each new article you read with previous information you have collected and summarized. Is this author supporting the claims of others? If not, can you determine why not? Are this article's findings significant, new information, or just careless and superficial thinking? Pay attention to the language of findings. Is the author claiming to have proved something or simply to have noticed a possible correlation between variables? You will need to be equally careful with your own language in making claims and recommendations on the basis of the author's information. Can you say that the study "shows" something conclusively? Are the findings "preliminary" or "conclusive"? If you use inaccurate language in your summary, those inaccuracies will inevitably get translated into your final report. Be careful with your note taking and summarizing. If you do them right, you will save yourself time and greatly improve the credibility of your writing.

4. *Organize* **all that information.** Do not be tempted to start downloading everything you find on your first search. Screen those sources through a further search, using more specific keywords. To manage those piles of information you're collecting, you will want to set up a document-management system, probably on your computer. You have already collected the information you will need to access each source later. Store that documentation in one word-processing file, for both on-line and print sources. You can cut and paste the citations from your library's on-line catalog.

Keep checking the relevance of the information against your outline, and weed out information that does not seem as relevant now as when you started the project. Be careful, though: Don't throw anything away until the project is finished.

If you have bothered to download some articles or citations, you might come around again to thinking that they *are* useful after all. So, stash them in a separate file called something like "Outtakes."

Documenting Information in Your First Draft

Once you start writing, you will need to carefully cite all the sources of the information and quotations that you use, so include your in-text citations in your first draft (or at least some form of notation that reminds you to include the citations later). Many engineering disciplines prefer that you place a parenthetical citation at the end of the sentence in which you have incorporated someone else's idea or words. In that citation style, taken from the American Psychological Association (APA), you put the author and date in parentheses:

> *One researcher has reported that air quality is declining in most U.S. cities (Turnbull 1998).*

There are basically three reasons to cite the source of your information. Here are examples of how to cite using the APA style for each of those three reasons

- Referencing ideas or facts found in the source:
  RDX has been used in a mixture with other explosives to produce bursting charges for bombs (Rizk 1996).
- Quoting directly from the source:
  Research has shown that RDX "has long been used in a mixture with other explosives" (Rizk 1996, 24).
- Giving credit to the author as part of the flow of your sentence:
  Rizk (1996) reviewed the uses of RDX and found they included being used in a mixture with other explosives.

Other styles use superscript numbering to indicate an in-text citation. In place of the parentheses, you place a superscript number that corresponds to the number of that source in the Reference List. So, the previous example would like like this:

> *RDX has been used in a mixture with other explosives to produce bursting charges for bombs.*[2]

The superscript 2 indicates that the full reference with all publication information is found as the second item in the reference list:

> Rizk, J.P. 1997, *Laboratory Investigation of Vadose Zone Characterization at the Pantex Plant, Amarillo, TX*, M.S. Thesis, University of Texas at Austin.

In this style, references are not arranged alphabetically but rather are numbered according to the sequence of their appearance in the text of the document.

Here are some guidelines for handling special cases of in-text citation.

- If your citation refers to material found in several sentences, place it in a topic sentence:
  A much different account of the goals of risk communication is found in the seminal work published by the National Research Council (1989).
- If your source does not name the author, give the name of the sponsoring or publishing organization:
  (National Cancer Institute 1993).
- When naming two or more sources in one place, separate them with semicolons:
  (Justin 1994; Skol 1972; Wiess 1986).

2.5 AVOIDING PLAGIARISM

What does the term "plagiarism mean to you? Do you associate it with cheating on exams? With copying someone else's paper? Do you associate it primarily with crabby professors who don't understand how busy you are or who don't have better things to think about? Well, cheating on an exam is "cheating," which is defined by most universities as any one of a number of ways of taking and submitting an exam improperly. "Plagiarism," on the other hand, is defined this way at the University of Texas at Austin:

> *"'Plagiarism' includes but is not limited to the appropriation, buying, receiving as a gift, or obtaining by any means someone else's work and then submitting that work for credit as if it were one's own"* (<u>Student Judicial Services</u>).

What that definition means is that you cannot pass off any part of anybody else's "work" as your own, even inadvertently. Because very often (in the workplace as well as in school) that work takes the form of "words," you have to be careful that the words you use in your papers and reports are your own. Most guidelines on avoiding plagiarism suggest making sure that you have not "lifted" more than 5 words in a row from another source.

Now, that's all fine, but what about engineering "words"? We all know that descriptions and specifications for products, components, or processes all have standard language, and you sometimes cannot vary the words and be precise. Here's a description, for instance, of a component of a particular type of septic system, written by Thomas H. Miller and found on the website of the Maryland Cooperative Extension:

> *"An alternative to the common drain field is the Seepage Pit (Dry Well). In this type, liquid flows to a pre-cast tank with sidewall holes, surrounded by gravel. (Older versions usually consist of a pit with open-jointed brick or stone walls.) Liquid seeps through the holes or joints to the surrounding soil"* (<u>Miller</u> *2004*).

Well, this description is available to anyone accessing this website; so, what's wrong with just using the description in your own paper, without any reference to the source? And, if you do think you should at least cite the source, what's wrong with leaving off the quotation marks? If you answered "nothing" to one or both of those questions, you need to learn why plagiarism matters. Plagiarism matters because it's a matter of accountability.

If you do not use quotation marks, even if you cite the source, you are essentially claiming those words as your own and so you are **responsible** for them. The information, for instance, that "older versions" of the Seepage Pit were made from "open-jointed brick or stone walls"—you had better be able to stand behind that information and explain it further to any inquiring mind, because you have just made yourself responsible for the truth and accuracy of it.

What if you appropriate this statement by McDonald's about its policy toward the environment:

> *"McDonald's has a long-standing commitment to environmental protection. Our restaurants around the world have innovative programs for recycling, resource conservation, and waste reduction"* (<u>McDonald's</u> *2005*).

If you use those words without attribution (citing the source) and without quotation marks, you are responsible for the truth and accuracy of this public-relations claim. You had better be able to write knowledgably and defend in public those 'innovative programs" that you have claimed McDonald's has. But if you use quotation marks and cite the source, you are saying, in essence, "this is what McDonald's claims," and then the context of your document can make clear whether you are supporting that claim (with more evidence), simply listing it as a matter of fact (McDonald's makes this claim), or disputing it. But you don't have to defend it.

How about this description by Boeing?

> "The newest members of the Boeing 737 family — the 737-600/-700/-800/-900 models — continue the 737's pre-eminence as the world's most popular and reliable commercial jet transport. The 737 family has won orders for more than 5,200 airplanes, which is more airplanes than The Boeing Company's biggest competitor has won for its entire product line since it began business" <u>(Boeing 2004)</u>.

If you don't cite the source or use quotation marks, you are claiming to have documentation of the stated fact that Boeing has "won orders for more than 5,200" of its 737 airplanes. If you don't have that documentation, you had better cite the source for that number: 5,200. But do you have to use quotation marks around this passage as well? After all, isn't most of the passage just factual information with no special phrasing?

Well, one test for whether to use quotation marks is this: Is the information considered general knowledge — something that every educated person should know or that as passed into the public domain, like, e.g., the saying "a stitch in time saves nine"? You do not need a source citation for that saying. And probably no quotation marks either. But is it common knowledge that Boeing has "won orders" for that many planes? Probably not. And what does it mean to "win orders"? Is that a term that might have one meaning for one company (items already shipped) and another meaning for another company (intent-to-order indicated)?

If you learn how to paraphrase and summarize information **in your own words**, you don't have to use quotation marks so much. Here is an example of appropriate paraphrase. The first passage is the original, and the second, in bold, is the paraphrase. All the information of the original is conveyed, without direct copying, and key phrases are quoted.

1. "Two studies by the Madhya Pradesh state government in the 1990s and three more in recent years by independent groups — the most notable conducted by the Greenpeace Research Laboratory at the University of Exeter in Britain — found severe groundwater pollution and attributed it to the Union Carbide plant's waste.

"The company rejects those conclusions.

"Spokesman Tom F. Sprick said the Connecticut-based firm trusts instead a 1997 survey by India's National Environmental Engineering Research Institute that judged the water to be untainted. The company's own consulting firm, Arthur D. Little, which oversaw the institute's study, warned, however, that its tests were not comprehensive and that the water may not be safe to drink."

2. **Although Union Carbide rejects the conclusions, five studies in the last fifteen years have found "severe groundwater pollution" in Bhopal and assigned**

responsibility to the American company. Two of the studies were conducted by the Indian state government, and the others (including the British-based Greenpeace Research Laboratory) concur. Union Carbide cites in its defense a 1997 study by India's National Environmental Engineering Research Institute, which was overseen by the company's own consulting firm, Arthur D. Little. Union Carbide spokesman Tom F. Sprick's defense was qualified by the consulting firm, however, who state that the testing was not comprehensive (McPhate, 2004).

[Moore, C., et al. 2006. Used by permission.]

Here is an example of inappropriate paraphrase. Again, the original is the first passage (#3). In the second passage (#4), key phrases are reused without quotation marks, and the structure of the paraphrase is lifted directly from the original. Even though the source is cited, this passage would be considered plagiarism because the writer is taking credit for the wording and structure of the original.

3. "The head of New Delhi's Bhopal disaster office, Ramesh Inder Singh, said the issue has languished for so long only because the legal nightmare that followed the disaster — more than 1 million claims were lodged — has kept the office paralyzed.

"This summer, New Delhi endorsed a lawsuit under way in New York brought by Bhopal victims against Union Carbide that seeks to compel the company to clean the site and pay damages to victims. The U.S. court had required India's permission to proceed with the case.

"Himanshu Rajan Sharma, an attorney for the plaintiffs, said he believes Union Carbide has judged the lives of poor people in distant countries to be expendable."

4. **Ramesh Inder Singh, the head of New Delhi's Bhopal disaster office, has said that the legal nightmare of having more than a million claims made against Union Carbide has caused the issue to remain unresolved. In a separate but related incident, the India government endorsed a New York court's decision to allow a lawsuit pursuing grievances against Union Carbide on behalf of the victims. The attorney for the plaintiffs said that he believes the company is guilty of judging poor people's lives as expendable (McPhate, 2004).**

[Moore, C., et al. 2006. Used by permission.]

I hope you can see by now that avoiding plagiarism is important in the workplace and for your career, as well as in school. When it comes to your education, here are two more reasons to avoid plagiarizing:

1. The whole point of your education is to help you teach yourself to think critically by gathering information, weighing it, and adding your own thoughts and conclusions. How can your professors help you do this without clearly knowing what is your conclusion and what is someone else's. You're just cheating yourself if you don't clearly make the distinction.

2. The only way you can show that you've learned something is by demonstrating which ideas and facts and results you used to come up with your own ideas and results. Even the most brilliant scientist or engineer is simply furthering work that has already been begun. So, make clear the foundation on which your research or study is built.

SUMMARY

- Any writing project requires intensive planning before you start a draft. You should find examples of the kind of document you need to produce in order to better understand what sorts of information you need to gather and how to frame it.
- There are many sources of information, published and unpublished, and many methods for gathering that information. Become familiar with using search engines, databases, and indexes. Identify the journals and publications cited most often in your field of engineering.
- Experts are good sources of information, but only if you interview them properly. Gather your questions ahead of time, and use body language during the interview to show interest.
- You will need to manage the large volume of information that you will probably collect. Keep track of published sources as you go along, since you will have to document those sources that you use in your writing or presentations.
- Not all sources of information are credible, especially on the Web. Evaluate information by examining the credibility of the sponsoring institution or group and the author.
- Learn to paraphrase and summarize the information you gather, and always cite the source of information that is not general knowledge.
- When in doubt about whether or not to use quotation marks around others' words, use the marks. And then, of course, cite the source of those words.

PROBLEMS

2.1. Assume you are writing a paper on global warming for an Environmental Engineering conference. On a scale of 1 (most credible) to 5, rate the following sources in terms of credibility with *your audience*. Look up these URLs and *justify your ratings*. Use the checklist at the end of Section 2.3.

http://www.greenpeace.org/
(Greenpeace's URL)

http://www.epa.gov/globalwarming/
(The EPA's page on global warming)

http://www.megastories.com/warming/bangla/intro.htm
(An article about the effect of global warming on Bangladesh)

www.skepticism.net/global_warming/
(Links to articles skeptical of environmentalists' claims)

http://www2.msstate.edu/~krreddy/glowar/glowar.html
(Global Warming International Center maintained at Mississippi State University)

2.2. Interview an engineer, either in the workplace or at your school or college. Choose someone you believe to be an expert in his/her field. Take notes on his or her language and demeanor. Write a description of the words, gestures, tone of voice and/or facts that tell you this person is an expert.

2.3. Begin to research a topic in your field by using an internet search engine. Record the exact words you use for each search. What words did you use to find the best source for writing a report on that topic? How many searches did you have to perform to find that source?

2.4. You work for an auto-industry research company and your job is to find out how many students in your classes drive cars to school and what kind of cars they drive. Design a short survey to capture that information. Use the survey in at least one of your classes.

2.5. Using the information you gathered in Problem 2.4, write a short report about the makes and models of cars students are driving and what percentage of students drive rather than use other transportation to school. Can you interpret and explain these findings?

2.6. Using the information you gathered in Problem 2.4, conduct a series of interviews with particular students to find out why they drive the cars they do (if they drive at all). Discuss in a short report your findings and your explanation of the interview results.

2.7. Read an article on a topic in a class you are taking (the article can either be assigned by your instructor or chosen by you). Summarize the article in 200 words or less.

2.8. Summarize the article you read for Problem 2.7 in 100 words or less. Did you have to leave out some important information?

2.9. If you felt you had to leave out important information by doing the summaries in Problem 2.7 and/or Problem 2.8, describe the audience(s) you think might really need that information you had to leave out. Would you characterize those people as decision-makers, advisors, or implementers?

FURTHER READING

Chicago Manual of Style, 15th ed., 2003. University of Chicago Press: Chicago.
 This is the mothership of style manuals. This big volume contains more than you'll ever want to know about manuscript preparation, documentation, and the publishing process, but it will also have definitive answers for you on all those niggling little questions about style (e.g., Do you write out the number "twelve" or use the Arabic numeral?). In particular, the "Grammar and Usage" and "Punctuation" sections are a terrific reference tool for engineers. Most discipline-specific style manuals in engineering are based on the *Chicago Manual*. See the "Documentation" section for a clear, albeit lengthy, demonstration of the differences between APA and MLA styles. Since you may have been taught the latter in high school, you will want to acquaint yourself with APA style, now that you are in a technical field.

Ford, J.E. 1995. *Teaching the Research Paper: From Theory to Practice, from Research to Writing*. Scarecrow Press: Metuchen, NJ, and London.
 This collection of papers is meant for teachers, but it contains some good advice for everyone on how to approach and do productive research. There is a nifty research model on page 141.

Harris, M. (2006). *Prentice Hall Reference Guide to Grammar and Usage*. 6th ed., Prentice Hall, New Jersey.

Lay, M. Wohlstrom, B; Rude, C; Selzer, J; and Selfe, C. 2000. *Technical Communication*, 2d. ed. Irwin: Chicago.
 This is one of the best of the many textbooks on all aspects of technical communication. One of the best chapters for engineering students is Chapter 14, "Reports for Decision-Makers."

Leahey, T.H., and Harris, R.J. 1997. *Learning and Cognition*, 4th ed. Prentice Hall: Upper Saddle River, NJ.
 This weighty tome is not meant to be read in its entirety by anyone who is not a cognitive scientist. But it is a surprisingly friendly presentation of everything you might want to know about how humans learn and understand and why we behave the way we do. The

most relevant chapter for engineers is the one entitled "Thinking," which offers good strategies for problem solving, reasoning, and decision making.

Moore, C.; Hart, H.; Carpenter, M.; and Randall, D. (2004-2006). *Professional Responsibility Learning Modules*, Cockrell School of Engineering, University of Texas at Austin. http://www.engr.utexas.edu/ethics/primeModules.cfm

Ruszkiewicz, J., Walker, J.R., and Pemberton, M.A. 2003. *Bookmarks: A Guide to Research and Writing*, 2d. ed. Longman: New York.

A much-needed guide to the entire process of conceiving and executing a research and writing process, this book takes you from defining initial scope to drafting and finalizing the project. The sections on evaluating sources and on understanding academic responsibility offer advice and information that are hard to find in other places. The book is easy to use, with helpful tabs and nifty illustrations.

Walker, J.R., and Taylor, T. 1998. *The Columbia Guide to Online Style*. Columbia University Press: New York.

This reference book organizes and categorizes the standards for citing electronic documents. It also includes sections on how to create documents electronically for print publication and how to format documents for on-line publication.

Organizing Ideas and Facts: Starting to Write

Objectives

By reading this chapter, you will learn the following:

- how to start the writing process;
- how to help yourself in the drafting process;
- to use writing to help yourself write;
- to use writing to clarify your engineering methodology.

3.1 INTRODUCTION

Just thinking about starting to write a document or put together a talk is scary. No matter how much you know (or think you know) about a subject or a situation, when you try to organize your thoughts and write them down in clear, logical paragraphs . . . well, you quickly realize what a crooked path your thoughts tend to follow. The way we usually think (the way our brains work) is quite different from the way we know we need to write. We think in images, keywords, or clusters of words at most. We do *not* think in sentences and paragraphs; we think *associatively*—one thought reminds us of something else, even if only because we experienced that something else right after the first thought. Thus, the first time you learned what an isosceles triangle is might be forever intertwined with a memory of your first kiss, which happened right after geometry class. So, writing is always an act of translation—translating into a primarily linear form the ideas and concepts that occur to us piecemeal in a recursive, associative, often visual pattern. Therefore, it is important to learn how to become better friends with the process of capturing your thoughts, organizing them, and writing them up.

3.2 STARTING TO WRITE

So how do we perform this act of translating our thoughts into writing? Not quickly. Writing (or putting together a presentation) is a process that happens over time and goes through several stages. Like most engineering projects, writing any professional document involves these steps (only two of which include what you probably think of as writing):

- Define the objectives for writing.
- Define the audience.
- Plan the type and format of writing (e.g., informal letter, formal proposal).
- Draft the document.
- Evaluate the document (with outside help).
- Revise the document.

If you substitute the word "project" for "writing" in this list, the word "plan" for "document," and the word "participants" for "audience," you would be describing any project-management procedure. It may be a while before you become a project manager, but even as an entry-level engineer, you will work with your team to clarify and find expression for the objectives given by upper management. Whether the desired result of a project is a new design, an improved process, or a list of requirements, the project manager will go through these six steps, although not necessarily in strict sequence. There can be lots of loops back between any two of these steps. The "draft" stage, for instance, results in a test document that you should ask other people to evaluate. If your evaluators cannot understand the purpose of the document (the bottom line, so to speak), you may need to go back to the "plan" stage and brainstorm about how to highlight the purpose better. You may even need to go back to the "define objectives" stage and determine whether you yourself are clear about the ultimate purpose of this project or document. It may sound pretty stupid to talk about a project manager or any experienced professional not being clear about purpose, but confusion and conflicting goals are common to large organizations with dozens of projects being planned simultaneously by hundreds of employees.

The most important point to remember about this list of steps is that *you cannot skip forward in the initial sequence*. For example, you cannot try to write a final draft before capturing your thoughts in a first draft. Even a professional writer cannot do that! You cannot write clearly if you do not know, for instance, to whom you are writing. Take a project to design a city's bus shelters as an example. If you are a consultant responding to a request for proposals published by the city's municipal transit authority, then the first document you write is a proposal to that authority, "selling" the excellence of your conceptual design of the bus shelters. You will probably stress the cost effectiveness of the design, since municipal authorities are responsible to city taxpayers. When you are hired (on the basis of your terrific proposal) to create the final design, you will probably help the transit authority "sell" the design to the public on the basis of how functional and aesthetically pleasing it is. In each case, what you highlight in the document will be somewhat different. So, you need to do some up-front work on defining the audience, purpose, and type of communication *before* you begin a draft of the document.

KEY IDEA: Remember: Writing is a *process*, a series of steps. Sometimes you will repeat steps, but do not try to skip any.

Once you have defined (or worked with others to define) your objectives for writing, the audience you are addressing, and the type of document needed, you are probably ready to take a stab at writing. Whether you feel ready or not, if your research is largely done, you are ready to begin a first draft of your document. Of course, you could convince yourself that your research is not done yet. You can *always* tell yourself that there is more relevant information out there, because there always is. The more you get to know about a subject, the more you realize how much more there is to know. This is the point where having a good definition of your purpose becomes critical for you. If you are writing a design report describing a new prosthetic device for a medical-devices company, you had better know a lot about other, similar devices. But you do not have to know everything about all prostheses ever invented.

So, how do you get started? What strategies can you use to move from researching to writing? In the case of a design proposal, of course, you have the design to start with. And that helps a lot. For the bus-shelter proposal, you can start by describing the design visually and then proceed to list its benefits and compatibility with the urban landscape. You are now well on your way. You can circle back later and write the introduction and supporting sections. So, strategy #1 is this:

1. **Start with the visual.** Create your graphics (designs, figures, and even tables) first. Then write text to describe and connect them together. You will not be starting at the beginning, but who cares? *This is only your first draft.*

Here are some other starting strategies. You can use any or all of them. They all involve using informal writing or speaking to help yourself work up to the first draft. The strategies are as follows:

2. **Start with an outline** that you generate any which way. You can make random lists of ideas, topics, or keywords; you can make doodles; you can talk to someone else and ask him or her to take notes as you speak. An outline is a tool for you to use; it can be as individual as you are. And no one else need ever see it. Some of us learned to outline by using numbers and letters in a linear sequence. Use that format only if it helps you get started writing. Many word processors have an outline feature, but make sure that you are not intimidated by the numbering; keep the outline as free form as possible until you are writing the first draft. Do not make the mistake of thinking you have to begin by writing about the first topic on the outline. You can start with any topic, because *this is only your first draft.*

3. **Start with the research summaries you produced.** This is the point at which you will thank yourself heartily for having taken the time to create paraphrases and summaries of the information you collected. You have already gotten some of *your* words on paper or screen. If you have typed your summaries, you can simply cut and paste them into your first draft. It does not matter which summary or topic you start writing about, because *this is only your first draft.*

4. **Start by freewriting.** This strategy is one of a number of ways to do what is also called *mind mapping* or *mind dumping.* The term *freewriting* was popularized by Peter Elbow in 1973 and 1981 to describe a process he refined for freeing your thoughts from the prison of your mind and getting them down on paper. The technique is simple, if odd, for most technical writers. You simply put your hands on the keyboard and type for 10 minutes without stopping. If you cannot think of what to write about your subject, you write whatever is in your head. Sooner or later, you get down some useful information, and you have the beginning of a document to work from. And, perhaps most importantly, you have begun the document and moved past the stage of doing nothing but procrastinating. It doesn't matter which section of the document you start writing and it doesn't matter how informal your language is at this point because this *is only your first draft.*

5. **Start by talking to someone about the subject matter.** Most of us find it easier to talk through our ideas than to write through them. Find a friend or colleague who will listen to your explanation of what you have to write about. When you sit down to write, you will simply be continuing this conversation. You can use this technique later on, too, when you get stuck at a particularly difficult place in your writing. Just walk away from the computer and find a person who will listen.

KEY IDEA: Learn to use the start-up strategy that works for you. Everyone has trouble beginning to write, and everyone needs their own strategy to get past the fear of the blank page/screen.

3.3 DRAFTING FOR MYSELF

OK, so let us say you have got some kind of outline for your writing project—or maybe just a series of lists and notes loosely tied together. Or you have talked to a good friend or colleague about your project. And you have created some tables and figures to represent data you generated or collected. In other words, you have tried several of the aforementioned strategies—but, nonetheless, you tremble with fear as you open your word processor. It is all very well and good to talk about steps in a process, defining your audience, and using various getting-started strategies, but at some point you just have to sit down and start writing; you have to start typing sentences and paragraphs that will hold together, get the message across, and be grammatically correct (in the end). Almost no one is immune to the terrors of the blank page or the blank screen. When you write, you create something whole (a proposal, say) out of fragments—bits of ideas, concepts, and facts. You are engaged in heavy labor.

E X P E R T S   S A Y . . .

Peter Elbow says that the main obstacle for most writers is "fear of wrongness" (1981, p. xvii). As long as we have not committed ourselves to words on screen, we can keep the illusion that we really know what we need to know; we just do not want to take the time to write it down. The fact is that you do not know what you know until you write it, even if the draft is rough. As Elbow says, "Clarity is not what we start out with but what we end up with" (1973).

It helps to remember that writing is a process, and the stages cannot be collapsed together or skipped in their initial sequence. This fact means that you have to be realistic in what you demand of yourself. Do not expect to write that report section just right on the first try. Tell yourself that you *will* do several drafts of the document, not because you are a weak writer, but because that is what all sensible writers do. Period. Here are some more suggestions for handling the process of writing:

- **Be good to yourself while you are writing**.   Make sure you have water, coffee, tea, or whatever you like close at hand. Take breaks regularly, but not when you are stuck. Find the most comfortable working chair you can, and make sure your keyboard and monitor are at the right heights. And try to figure out when is the optimal time for you to write. First thing in the morning? Very late at night? There are no rules about "when" to write; just begin earlier than the night before an assignment is due.
- **Make the process less solitary by getting as much feedback as possible**.   Ask friends or colleagues to read drafts of your document (and agree to do the same for them). Or make a deal with a friend or colleague that when you are stuck, you can call him or her and chat about what you are stuck on (e.g., "How do I explain thermal dynamics to an audience who has never studied engineering?").
- **Do not criticize your writing as you write**.   Save the criticism for the final revision stage. Especially when you are drafting, stay upbeat and positive about your own writing abilities: "I can do this; I know I can." Professional writers spend a lot of time talking to each other about their individual strategies for staying cheerful

and confident. It may seem faster to edit as you draft, but in the end this practice will drag you down and make you write slower and slower. And anyway, how can you be a creator (first draft) and a critic (revised draft) at the same time? That is like trying to tell a joke when you are angry.

- **If you are collaborating on a document, stay in close touch with your collaborators.** Use e-mail, phone calls, or informal meetings to help you iron out problems and get past sticky spots. Be grateful for collaboration; the hardest thing for most of us about writing is that it is often so solitary. Of course, even when collaborating, you must spend some time writing on your own. But show your work to collaborators as soon as feasible, and get that all-important feedback.

3.4 USING WRITING TO THINK CLEARLY

Lots of people have said something like, "You do not know what you know until you can write it down clearly." At least some of the time, however, the reason it is difficult to write is that *we are not sure exactly what we want to say*. We may have thought about the concepts and purpose of a document we have to write, but as long as those thoughts stay locked in our head, they remain free-floating, vague, ghostly. Let us take as an example a passage of writing from the introduction of a paper on trying to control ozone in the atmosphere and work backwards into the author's original thoughts on the subject:

> Controlling ground-level ozone is one of the most significant environmental challenges facing the United States. Ozone is a major component of photochemical air pollution, or smog, and causes harm to both human health and the environment. Ozone is formed by atmospheric reactions of volatile organic compounds (VOCs) and nitrogen oxides (NO_x) in the presence of sunlight. For more than two decades, emission controls on VOCs and NO_x have been applied, yet ozone concentrations remain at unhealthy levels throughout much of the United States. There are many reasons that the ozone problem remains relatively intractable, but a critical element is the fact that not all VOC and NO_x emission reductions lead to equivalent reductions in ozone formation. *When* and *where* the emissions of VOCs and NO_x occur play a critical role in determining ozone formation rates.

The author has done a lot of research on how ozone forms in earth's close atmosphere and why it is potentially harmful in larger amounts. She also has researched various options for controlling ozone, including controlling emissions of harmful precursors of ozone (VOCs and NO_x), and the ineffectiveness of recent regulations in limiting the amount of ozone produced by these chemicals. So, we can imagine that before she tries to write the introductory paragraph, she lets her thoughts bring to the surface the most important words that keep recurring: "ozone," "controlling," "problem," "emission reductions," and "ozone formation." She jots those words down and starts to construct her introductory paragraph out of them. She also knows that the history of these attempts to control ozone is important, so she inserts a time element: "for more than two decades." And then she realizes that much of her research has involved looking at maps that plot the occurrence

KEY IDEA: You can use writing for *yourself*, to ease the transition from thought to sentence, from mental picture to paragraph.

of ozone over certain metropolitan areas, and that what is critical to understanding ozone formation is the spatial dimension of how and where harmful emissions are formed. So she needs to use spatial as well as time-oriented words to describe the process: "where" and "when."

Before long, she can construct sentences using those recurring words and string them together using transitions and organizing patterns such as chronology. She is well on her way to writing a clear, readable introductory paragraph, because she has translated her thoughts into straightforward, clear words and phrases. Remember that you can use writing—outlines, jottings, and phrases—to ease the transition from thought to sentence, from mental picture to paragraph. Use informal writing for *yourself*—to help translate your thoughts into writing.

E X P E R T S   S A Y . . .

Peter Elbow says that freewriting can work to get you started *and* to help your writing flow, because stopping and starting (our usual writing method) creates fragmented writing. Sentences that are freewritten "hang together," even though they still need editing, they have usually captured a whole thought (1998, p. 16).

Engineers can use freewriting to work their way out of hard-to-explain passages and to capture thoughts for the more imaginative types of engineering documents, such as proposals.

3.5 USING WRITING TO SOLVE PROBLEMS

KEY IDEA: Thinking on paper allows you to sort, categorize, and synthesize multiple types of data and information.

Writing can actually help you solve problems with your engineering work, especially problems with your methodology and with understanding the full significance of your work. Let us say you are a modeling engineer who models underground reservoirs for oil and gas exploration companies. You use particular models, generated by different software, to solve different issues for clients: size of reservoir, depth, rock and soil formation, action of subsurface water, likely presence or absence of oil and gas over time, etc. You start projects by collecting a lot of field and other data, and then you determine the parameters of the particular model and run it, probably many times. From the moment you start collecting the data that will be inputs to the model, you are involved in analysis that could benefit from being written down. What sort of data are the best as inputs for a particular model? Thinking on paper allows you to sort, categorize, and synthesize multiple types of data on the basis of past experience and of your engineering analysis of the particular situation. In addition, if you write up your thoughts and insights about the data collection, you will already have part of a methodology section written for the final report you have to present to the client. In fact, since these sorts of insights about the data collection are necessary for you to formulate the precise goals of your modeling work for this client, you can now also write a draft of the introduction to the final report. Understanding your overall goals is an excellent way of staying on track with your modeling and number-crunching work. And writing helps you understand (just as understanding helps you write).

Writing can benefit your engineering work throughout the entire project. As often happens, for instance, you may not know whether the model will give you the information you want until you start running it. If the model produces outputs that don't make sense, it could be that you have provided the wrong inputs and need to rethink your methodology. If you have already drafted a data section, it is much easier to pinpoint the data or the calculations that may be faulty or inappropriate. You can then revise your inputs much more easily. Or it may be that your inputs *are* appropriate, but the results (outputs) are very difficult to interpret. Documenting the model's results will help you see connections between them and may lead to the sort of "Aha!" insight in which you suddenly see *why* what happened, happened.

Writing as you perform your engineering work allows you not only to break down the writing process into manageable stages, but also to use what you write to identify mistakes in your approach or in your interpretation. You would not try to keep a string of equations in your head, and neither should you try to keep a string of thoughts in your head. When you interleave *documenting* the work and *doing* the work, you allow each type of activity to improve the other. This approach means writing in increments, using what Herbert Michaelson calls the "Incremental Method" (1990). The trick is to remember that you are using writing to solve problems rather than to impress a reader. Most likely, this method will lead to a draft of the final written product, but that draft will need reorganization, editing, and probably completion. Until that revision stage, however, you are drafting for *yourself* (and perhaps for a trusted content reviewer). You are using writing to help yourself think, analyze, and solve problems.

SUMMARY

- Starting a writing project is quite scary. Everyone should use a start-up strategy that works for him or her. The basic trick is to remember to smother your inner critic at the beginning and bring him or her back into play only when you have finished drafting the document.
- Some writing is done entirely for yourself (lists, initial outlines, etc.); no one needs to see the way you used these forms of writing to get to the stage of actually writing the document. And no one need ever see your drafts, unless you want a review. Writing is an aid for your thinking process, and it can even be combined with that other aid, doodling.
- Writing is a form of problem solving. You can use writing to help plan the direction of your engineering work as well as to frame the document you have to write. Documenting your work *as* you work improves your engineering analysis.

PROBLEMS

3.1. Try one of the start-up strategies described in this chapter. Use the strategy to organize your thoughts and write a first draft of any writing assignment you have currently. Keep a record of how long it takes you to begin writing the first draft (completing at least one paragraph) and then how long it takes to complete the draft. Also record how comfortable you feel with that strategy. Hand in these notes to your instructor.

3.2. If your chosen strategy in Problem 3.1 did not work, then for your own benefit, try another strategy later for another assignment. Determine whether that one is more comfortable for you and prepares you better for writing the draft. Explain why you are or are not better prepared.

3.3. Think back to a recent writing project. Did you follow the steps listed in Section 3.1? If you skipped any, do you think your final product was affected negatively? Write up your thoughts on this matter.

3.4. How might an engineer use freewriting to help himself or herself get started writing a report on how to improve a particular work-flow process?

3.5. You have been asked to write a report requesting funds for studying solutions to the parking problem at your school, university, or place of work. The report will be written to the school president or the CEO. Brainstorm with another student and produce a list of all the topics you will have to address in such a report.

3.6. Partner with another student to begin writing the report outlined in Problem 3.5. List the kinds of information you will have to gather and decide who will gather which information. On what basis did you decide how to divide up the research?

3.7. For the report requested in Problem 3.5, talk to a sampling of students about how much time they take to get to work or school in the morning. After the interviews, *first*, plot the results in a graph or chart. *Then*, write a paragraph explaining the visual. Answer these questions:

How long did it take to write that paragraph? Do you think the writing went faster because you had already created the visual?

3.8. The first draft of any document is usually written for whom?

(a) The author's boss
(b) The primary audience
(c) The author

FURTHER READING

Elbow, P. 1998. *Writing with Power: Techniques for Mastering the Writing Process*, 2d. ed. Oxford University Press: New York.

——1973. *Writing Without Teachers.* Oxford University Press: New York.

These are two of the most influential books on the process of writing ever written. They broke new ground by focusing on the *way* we write rather than just *what* we write. Engineers may not want to try all the exercises in these books, but they will benefit from listening to Elbow's careful articulation of the distinction between the critical and the creative functions of our brain.

Michaelson, H.B. 1990. *How to Write and Publish Engineering Papers and Reports*, 3d. ed. Oryx Press: Phoenix, AZ.
In this handy and brief book, Herbert Michaelson outlines with admirable clarity the organization of engineering papers and reports. The focus is on journal papers, but this is not to say that the book is academic in language or tone. Michaelson's goal is to encourage engineers, wherever they are working, to publish more. Chapter 5, "How to Use the Incremental Method," makes a good case for writing as you proceed with a project; Michaelson points out that publication will be much faster when this technique is used.

CHAPTER 4

Writing: Taking Control

Objectives

By reading this chapter, you will learn the following:

- to use a readable style for your technical documents;
- to recognize passive-voice constructions and use them appropriately;
- how to construct logical, organized paragraphs;
- to recognize sentence fragments and avoid them;
- how to choose the correct spelling of commonly misspelled words;
- how to choose the proper verb tense;
- how to control sentences through punctuation.

4.1 WRITING FOR CONTROL COMES *AFTER* DRAFTING FOR CONTENT

When you draft a document, you are trying to get your thoughts down on paper or screen. You are trying to determine the content of what you need to say, the data you need to present, and the arguments you need to make. You should not worry during the drafting phase about grammar, punctuation, and all those other things that the word "writing" may mean to you. If you worry about them too much, you will slow yourself down. You want your draft to be as complete as possible, and *then* you can go back and correct it for writing errors.

When you have completed a good first draft—that is, when you have written something about all the items on your outline or you just believe that you have covered the territory of what you have to say or present—then you are ready to focus on the quality of your writing. If you have already cognitively absorbed the guidelines presented in this chapter, you will of course have a better quality draft than if you know very little about crafting clear sentences and paragraphs. But either way, be easy on yourself and remember that you will still need to get external feedback before you deliver a final document.

4.2 READABLE STYLE

The style of technical writing has long been thought to be different from the styles of other types of writing, such as creative writing, essay writing, and journalism. And it *is*, although not as different as many people think. As a technical writer, you still have to grab a reader's attention, make logical arguments, and carry the reader along in a flow of words that gets your message across. The difference is that technical writing focuses less on you, the writer, and more on the ideas or facts you are presenting. As with all writing, you are responsible for helping the reader understand the message, but the message in this case often involves complex data and information.

Even in the most technical documents, you will want to be as concrete as possible so that the reader does not have to work too hard to learn the new information you are offering. Research shows that we read at two levels of comprehension as we process information: superficial processing and deeper processing. In the first type, we simply try to retain facts; in the second, we attempt to comprehend relationships among the facts (Der Meer and Hoffmann, 1987, p. 119). In addition, we try to remember new facts by seeing whether they link up with what we already know. If the link is clear (for example, if the author uses transitions and sequences information logically), we are well on our way to understanding the new information because we can begin with what we already know. If new and important information is highlighted by the author, we get help in understanding the content.

RESEARCH SHOWS...

Cognitive psychologists study how people read and understand information. Their studies have shown that these elements enhance the processing of information for the reader:

- **visual imagery**
- **examples**
- **comparisons**
- **descriptions**
- **questions**

(Der Meer and Hoffmann, 1987)

Let us look at an example of clear technical writing. Even if you know nothing about phase-sensitive X-ray imaging, you will probably easily understand this passage from *Physics Today* (July 2000, p. 23):

> Clinical and biological studies stand particularly well poised to benefit from the development of phase-sensitive techniques. Absorption contrast works well in distinguishing between hard and soft tissue: Heavier elements—like calcium in bones and teeth—have a much higher absorption cross section than the lighter elements that constitute soft tissues. However, in many clinical situations, such as mammography, there is a need to distinguish between different kinds of soft tissue—between tumors and normal tissue, for instance. Because the absorption is small to begin with, and differences in density and composition are slight, standard x-ray imaging is not as successful at this task.

Readable style is generally grammatically correct, but it is also a lot more than just that. In fact, grammatical correctness may be less important overall than presenting information logically and leading the reader carefully along the trail of your thought. To lead your readers, you need to prune your writing of deadwood, make the sequence of your thought explicit, avoid ambiguity, and keep the reader focused on the most important points. The keys to good writing, outlined in the upcoming box, are just as important in technical writing as in any other type of writing. They are not so much rules of grammar as strategies for writing clearly and concisely and for helping your reader remember the information.

KEY IDEA: Your job as an engineering writer is to help your reader not only to understand your information, but also to remember it.

4.2.1 Passive and Active Voice: What Is All the Fuss?

Perhaps the largest difference between technical and other kinds of writing is the increased use of passive voice in the former. The word "voice," as applied to writing, just means how you "sound" on the page instead of at the lectern. Your writing has a voice because the reader, in some not-very-well understood way, "hears" the words on your page in his or her head. Voice in writing is different than in talking, because the reader can slow down and "listen" again to your words. So you have to make sure that your "tone" is exactly what you intend. Both passive voice and active voice can strike the wrong tone when used inappropriately.

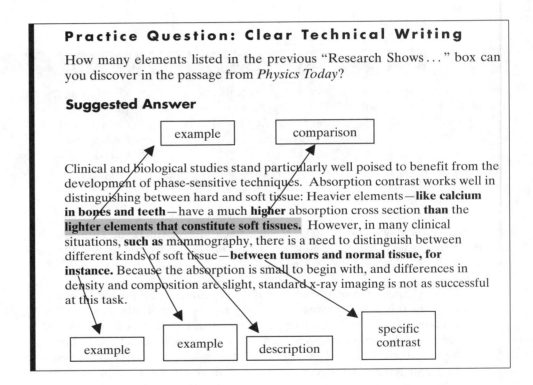

Practice Question: Clear Technical Writing

How many elements listed in the previous "Research Shows..." box can you discover in the passage from *Physics Today*?

Suggested Answer

example   comparison

Clinical and biological studies stand particularly well poised to benefit from the development of phase-sensitive techniques. Absorption contrast works well in distinguishing between hard and soft tissue: Heavier elements—**like calcium in bones and teeth**—have a much **higher** absorption cross section **than** the **lighter elements that constitute soft tissues.** However, in many clinical situations, **such as** mammography, there is a need to distinguish between different kinds of soft tissue—**between tumors and normal tissue, for instance.** Because the absorption is small to begin with, and differences in density and composition are slight, standard x-ray imaging is not as successful at this task.

example   example   description   specific contrast

KEY IDEA

Keys to Powerful Writing

There are five keys to good writing. The last three contribute to both the big picture (paragraphs and sections) and the smaller picture (individual sentences). The first two keys keep your sentences clear, concise, and communicative.

prune your writing
control passive voice
maintain logical linkage
maintain focus
maintain parallelism

What is the difference between passive and active voice, and why should you care? Well, compare these two sentences:

I threw the ball.
The ball was thrown by me.

The first is in active voice, the second in passive. Say these sentences aloud to yourself. The first is the way we would describe the event when talking; the second kind of sentence usually occurs only in writing, not in speaking. The agent (or doer) of the action is clear in the first sentence; the doer may not be so clear in the second version, especially (as often happens with passive sentences) if the doer is left off entirely, as in "The ball was thrown."

When to use which voice is an ongoing issue for technical writers, who often present information in which the doer of the action is not as important as what was done (the action) or what it was done to (the object). In this sentence, for instance — "The material was heated to a temperature of 100°C," the reader does not care who heated up the water sample, so naming the doer of the action would simply waste the reader's time. We do not need to use the active-voice construction, "The technician heated the material to a temperature of 100°C," because we do not need to know who is the agent (the technician) of the action of heating. Teachers of creative and essay writing encourage students to use only active voice, because active-voice sentences are usually clearer and more direct. But technical writers, including you, must decide what they want to emphasize before they can construct a good sentence and use the appropriate voice.

When to Use Which Voice?

Active Voice	Passive Voice
• **To emphasize the agent of the action:**	• **To emphasize the object:**
Previous researchers established the relationship between chemical qualities and model parameters.	The relationship between chemical qualities and model parameters was established by previous researchers.
• **To be concise and direct:**	
JMC Consulting recommends that the agency adopt a risk communication plan.	The water sample was heated to a temperature of 100°C.

4.2.2 Wordiness: Learn to Prune

One reason that technical writers use a lot of passive voice may be that they want to avoid using the personal pronouns "I" and "we." Some engineering journals and many engineering faculty do not allow those words, and so the doer of the action cannot be easily named. The problem is that the reader may get confused about who did what, especially when it comes to attributing authorship or research effort. If you write, "This experiment demonstrated that chlorine dioxide is a viable water-treatment method," you had best make sure that the context makes clear whose experiment it was. Yours? Another researcher's? Check with your professor about using "I" and "we" and use substitutes, such as "the author" (meaning you) or "the

CE 333T research team," to indicate that the experimental work is yours. In any case, do not use the informal "you" in documents.

A big complaint about technical writing is its wordiness. A sentence like the following is typical of much technical writing:

> An important conclusion extracted from the result is that there is a correlation between the mixing length and the volumetric flow.

The sentence is not terribly long, but it wastes words. And those "dead" words get in the way of our quick and easy understanding of the meaning of the sentence. The following is a better way to write it:

> The result demonstrates a correlation between mixing length and volumetric flow.

The revised sentence has 11 words, almost half the length of the original, which has 21 words. The second sentence relates the important conclusion without wasting time announcing that we are about to hear an important conclusion. If the word "important" seems important, there are other ways of declaring importance to a reader, such as italicizing the sentence or making sure that the sentence begins or ends a paragraph.

Our writing models are often not very good, and we pick up habits of using unnecessary words and legalisms. Also, the way we first write something when we are drafting is often the long way around explaining it. *Pruning* is the process of going through your own draft and removing the deadwood—the words that really do not add anything to the meaning you are trying to convey. Just as pruning a tree makes it stronger and eventually more beautiful, pruning your sentences will strengthen what you are trying to convey. See Table 4.1 for some examples of how to prune common deadwood phrases.

Table 4.1 Words and Phrases that Can often be Pruned

Deadwood	Pruned wood
different aspects	aspects
make a measurement	measure
give consideration to	consider
relative to	about *or* of
it should be noted that . . .	note that . . .
regardless of the fact that	although
due to the fact that	because
subsequent to	after
in the course of	during

Practice Question: Pruning

How many words can you remove from this sentence without changing the meaning?

> **The following report outlines the different aspects relating to the current feasibility of constructing a magnetically levitated bullet train in Central Texas.**

Suggested Answer

> This
>
> ~~The following~~ report outlines ~~the different aspects relating to~~ the *current* feasibility of constructing a magnetically levitated bullet train in Central Texas.

> The word "current" is also a candidate for pruning; it depends on whether you want to emphasize the report's timeliness (perhaps in comparison with another report from last year) or not. Normally, feasibility studies cover only current feasibility.

But do not prune just to make sentences shorter! Shorter sentences are not necessarily better than longer sentences; they may in fact leave out important information. Or they may actually hide meaning as the reader is forced to bump along, stopping at every period and starting up again. Consider this string of short sentences:

> This report is a final report. It presents final results for the design of a wastewater treatment plant. The design is for the process sequence, piping, flow distribution, and hydraulics. It includes the pump station and biological-treatment units. It also includes the disinfection and sludge-handling processes for the plant.

Notice how confusing those sentences become as they keep stopping and starting. What does "it" refer to at the beginning of three of the sentences: the report, the design, or some process? "It" is a very small word that does not mean anything by itself. Not only are the sentences choppy, but also we have to keep reading backwards to figure out what "it" means each time. Here, the best solution is to combine some of these sentences, not to shorten them:

> This final report presents the design of a wastewater treatment plant, including the process sequence, piping, flow distribution, and hydraulics. **The design** also includes the pump station, the biological-treatment units, and the disinfection and sludge-handling processes for the plant.

We have lengthened the sentences in this passage, but by combining some sentences, we have actually *shortened the passage*. Productive pruning can be a two-step process in which you actually lengthen a sentence in order to strengthen the passage. Again, pruning is not better for its own sake, but only when it removes deadwood. Here, some of the deadwood is the meaningless word "it" that keeps cropping up, which leads us to the next major problem with much technical writing: empty pronouns.

Practice Question: Combining Sentences

Consider the following sentences:

> **The population explosion in and around the Austin area has proved to be problematic regarding water-drainage structures. It has particularly affected the Brushy Creek area in southern Williamson County.**
> - Does combining these sentences improve their readability?
> - What problems does combining the sentences solve?

Suggested Answer

> **The population explosion in and around the Austin area, particularly in the Brushy Creek area in southern Williamson County, has proved to be problematic regarding water-drainage structures.**

By taking the geographic details of the second sentence and inserting them into the first sentence, we can keep "population explosion" as the subject of one clear sentence with one predicate ("has proved to be"). Longer sentences may produce a shorter passage! Of course, we can prune this sentence even more—see the next Practice Question.

4.2.3 Empty Pronouns

Isn't it annoying to read sloppy sentences that use the word "this" instead of a word with some meaning, such as the following?

> In the United States, the chemical and petroleum industries consume eight million tons of hydrogen annually. **This** is produced from natural gas by a process called steam reforming that extracts hydrogen from hydrocarbons. **This** can occur by one of two methods.

KEY IDEA: Never use "this" by itself at the beginning of a sentence. Always use "this" to modify a noun—for example, "this *process*" or "this *chemical reaction*."

When you use "this" without any noun attached to it, especially at the beginning of a sentence, you are cheating the reader. Guessing what the author means by "this" is a game that we readers would rather not play. In the second sentence of the foregoing example, we would like to be sure that the word "this" refers to hydrogen. (Actually, it refers to the eight million tons of hydrogen consumed by the U.S. chemical and petroleum industries.) Hydrogen, however, is not particularly emphasized in the previous sentence; it is tucked into a prepositional phrase ("of hydrogen"). So why not say, "**This hydrogen** is produced . . . "? And what does "this" refer to in the third sentence? As readers, we prefer the clarity of "**This process** can occur by one of two methods," or better yet, "**Steam reforming** can occur. . . . "

Use of "it" at the beginning of a sentence can also be a problem. Generally, when we begin a sentence with "it," we are delaying getting to the point. For example, consider the sentence, "It is the case that hydrogen occurs naturally." We can prune away the first five words without changing the meaning of that sentence! "It" has its place, but watch out for cases where you are using "it" because you do not know exactly what you want to say or because you are being unthinkingly wordy.

4.2.4 Noun Compression: Let Some Air in

Another barrier for readers of technical writing is long noun phrases, such as *process design revamp project* or *multireservoir water resources systems*. Even if you have seen these phrases before, it is difficult to know exactly what they mean. Here is a sentence that invites you to skip right over it:

> The scope of work provides technical service support to continue to improve the quality of the company's Advanced Science and Technology Plan and **develop a gap-analysis-based, site-specific Integrated Technology Plan** that includes Plant Directed Research and Development Thrusts.

Problems abound with this sentence, including the prevalence of strings of capitalized nouns that mean nothing to a reader unfamiliar with these "plans" and "thrusts." But even the noncapitalized noun phrases are problematic. What does "gap-analysis-based" mean? "Based" is actually an adjective, not a noun, but it does not help us much to discover the meaning. Does the phrase mean an analysis based on identifying the gaps in the current plan? If so, that part of the sentence should read this way:

> develop **a site-specific Integrated Technology Plan that fills in the gaps in the current plan**

If we insert this more understandable phrase into the original sentence, however, we will make the sentence even longer, so perhaps in this case we should split the sentence in two:

> The scope of work provides technical service support to continue to improve the quality of the company's Advanced Science and Technology Plan. **The scope should also** develop a site-specific Integrated Technology Plan that fills in the gaps in the current plan and includes Plant Directed Research and Development Thrusts.

The new passage is longer than the original, but it is much clearer. Of course, we can always work on pruning it. For example, do you think "technical service support" means anything more than "technical support"?

Noun phrases clog up our writing because they try to stuff so much meaning into a small place that there is no room to understand the meaning. Sometimes you have to add words to clarify the relationship of words to each other. Look at this sentence, for instance:

> This report on statewide outdoor burning describes the **activity data collection process**.

Surely some of the nouns in the indicated phrase are being made to act as adjectives, but we are uncertain about what is referring to what. Adding prepositions and pronouns may help to clarify compressed noun phrases. What kind of activity are we talking about? The incidents of outdoor burning? In that case, a better version would look like this:

> This report describes **the process of collecting data on incidents of** outdoor burning statewide.

KEY IDEA: Shorter is not always better. Clearer *is* always better.

The solution is to add some prepositions that indicate the relationship of certain words to each other. Here, too, the new sentence is *longer* than the original, but the changes *add* to the meaning rather than detract from it.

Practice Question: More Pruning

We previously produced the following sentence by combining the original two sentences:

> **The population explosion in and around the Austin area, particularly in the Brushy Creek area in southern Williamson County, has proved to be problematic regarding water-drainage structures.**

Can you make it even more concise (without losing meaning) by pruning it some more?

Suggested Answer

Here is the sentence with changes shown and then with the changes incorporated.

> **The population explosion in and around the Austin area, particularly in the Brushy Creek area in southern Williamson County, has ~~proved to be~~ *created* problematics ~~regarding~~ *for* water-drainage structures.**

> **The population explosion in and around the Austin area, particularly in the Brushy Creek area in southern Williamson County, has created problems for water-drainage structures.**

The changes have made this sentence both shorter and clearer.

4.3 CONSTRUCTING POWERFUL PARAGRAPHS

Here are the three keys to writing clear paragraphs and sections of a document:

- maintain logical linkage;
- maintain focus;
- maintain parallelism.

Maintaining links, focus, and parallelism is a matter of style and emphasis; using these keys requires that you think hard about exactly which bits of information and data are most important.

4.3.1 Maintain Logical Linkage

Good technical writing is not list making. We do not learn anything from lists unless we already know the context for all the items on the list. When you go to the grocery store, you already know that you are buying items for dinner, for instance. But you probably do not write down, "Dinner for four people." And yet knowing that information is critical to understanding how much of each item to buy. You cannot really use your grocery list—e.g., chicken, lettuce, tomatoes—without understanding that unwritten context. Good writing helps the reader understand the context by making explicit connections between ideas or pieces of information. These connections are like signposts that say "A is like B" (comparison), "A caused B" (cause and effect), "A is different from B" (contrast), or "A is more important than B" (degree of control). These signposts use conventional words and phrases such as "similarly," "consequently," and "on the other hand" to show the reader where your thought is heading.

Practice Question: Maintaining Linkage

Here is a paragraph from the Work Remaining section of a progress report. Do you know in what direction the thought is headed in each sentence? If not, can you diagnose what is missing from the paragraph?

> **The evaluation of diamond film based on the final criterion will be completed when the analysis of its qualities is received from the Institute for Advanced Technology (IAT). A request for the resource files on diamond film will be submitted to Bolton Manufacturing. An additional comparison of the results received from IAT will be made with information provided by Bolton. The evaluation of diamond dust will be used for the comparison of the other nanocrystalline coatings.**

Suggested Answer

There are no links between these sentences, no words that establish a conceptual connection. Every sentence seems to start the thought process all over again.

Let us take a closer look at the paragraph in the foregoing practice question to understand what has gone wrong:

> The evaluation of diamond film based on the final criterion will be completed when the analysis of its qualities is received from the Institute for Advanced Technology (IAT). A request for the resource files on diamond film will be submitted to Bolton Manufacturing. An additional comparison of the results received from IAT will be made with information provided by Bolton. The evaluation of diamond dust will be used for the comparison of the other nanocrystalline coatings.

The sentences generally describe what will happen, but we do not know the connection between each of these promises. Are they listed in chronological order? Will the most important steps be taken care of first? Do certain steps depend on others and require that the others be accomplished first? What does Bolton Manufacturing have to do before the author can accomplish certain steps? The paragraph is simply a list of unrelated steps and reveals almost nothing about the sequence of this process (except for the word "when" in the first sentence).

KEY IDEA: Your job as an engineering writer is to provide *connections* between the bits of information you are presenting.

Part of the problem with this paragraph is overuse of the passive voice; we are not sure who is doing what when. But even if we retain most of the passive-voice constructions, we can greatly improve the readability of the paragraph by adding links. Lets us look at a revised version of the paragraph. Why is this revision so much easier to understand?

> The evaluation of diamond film based on the final criterion will be completed when the analysis of its qualities is received from the Institute for Advanced Technology (IAT). **In addition**, a request for the resource files on diamond film will be submitted to Bolton Manufacturing, **so that we may compare** the results received from IAT with information provided by Bolton. **Once our evaluation of diamond film is completed**, we can compare diamond film with the other nanocrystalline coatings.

The paragraph now includes some links between sentences and between the concepts of the second, long sentence. We know a lot more about where each sentence is going. What made the difference in this paragraph? Transitional phrases.

KEY IDEA

Transitions establish strong links between these elements:

parts of a sentence;
whole sentences;
whole paragraphs.

Practice Question: Transitions

How many words or phrases can you think of that would fit in the blank at the beginning of the second of the following two sentences?

> *The project manager will be in Japan for the next two weeks. _____, the project engineers will remain in town at the branch office.*

Suggested Answer
Any of these words would make sense as a link for the two sentences:

However
Nevertheless
Therefore
Unfortunately

There are other possible links as well. Each one produces a different meaning. You have to provide the *right* link to the reader to convey the right meaning (the one you intend).

When you consider these two sentences and the relationship between them, you realize that lots of different relationships are possible:

> The project manager will be in Japan for the next two weeks. _____, the project engineers will remain in town at the branch office.

Each of those relationships would be indicated by a different word or phrase in that blank space at the beginning of the second sentence, such as "On the other hand," "Therefore," or "However." Each of these transitions imparts a different meaning to the sentences. Which is the right meaning? As the writer, you want to establish control and guide your readers to the right meaning. You may want to emphasize contrast ("However") in order to make clear the difference between the situations of managers and nonmanagers. Or you may want to emphasize cause and effect ("Therefore") in order to explain why the engineers cannot travel even locally during the next two weeks. **If you do not establish the correct connection between sentences and phrases, your reader will attempt to establish the connection anyway.** And that might be the wrong connection or meaning. For example, do you want your readers to assume that the missing word in the blank is "Unfortunately"? (This example also shows that sometimes two sentences are better constructed as one: Especially if you use "However," you would want to make one sentence, with a semicolon before and a comma after: "; however,".)

Table 4.2 lists some useful transitional words and phrases. Note, however, that this list does not begin to cover all the varieties of transitional words and phrases.

Table 4.2 Partial List of Transitions and the Relationships They Indicate

Transitional Word or Phrase	Indicated Relationship of Phrases, Sentences, or Paragraphs
similarly	Comparison
therefore; consequently; as a result	Cause and effect
but; however	Contrast
first; second; finally	Sequence
and; also	Addition
for example	Example
partly; the other half	Classification
upwind; downwind; to the left; to the right; in back; in front	Spatial relationship
most importantly; secondarily	Relative importance

KEY IDEA: The right transitions guide the reader down the right path.

Comparison and Contrast: a common pattern of organization for paragraphs and even whole sections of a document.

Transitions also create organization. In a paragraph such as the next example, the transitional phrases establish the organizing pattern—comparison and contrast—that helps us make quick sense of the writing. In this paragraph the author (in *Physics Today*, August 2002, p. 17) is talking about how a team of researchers is attempting to quantify the amount and distribution of water ice on Mars:

> For the ice's depth distribution, the team considered **two simple cases. In one**, the ice-bearing soil extends right to the surface; **in the other**, a dry layer of variable thickness tops the ice-containing layer. Thermal and epithermal neutrons, thanks to their **different behavior in the two cases**, provide the discrimination.

We can see classification also at work here ("two" cases, "the other").

You will notice that we have already looked at several of these patterns, since making clear the precise relationship among facts, ideas, or events is the main goal of much technical writing. In addition, the patterns often overlap within the same paragraph or section of a document. Which patterns are operating in this paragraph from an article on X-ray imaging (*Physics Today*, July 2000, p. 26)?

> A monochromator throws much of the x-ray flux away, and when used with an x-ray tube source, would make the time needed to record an image unacceptably long for clinical uses. The ability to use the polychromatic output of an x-ray tube is **therefore** important for potential clinical applications. The large beam divergence from such a source has the **additional** advantage of providing magnification, with a corresponding increase in spatial resolution. The spreading beam **also** allows the imaging of large areas in a single step.

Notice that in this paragraph, the transitions are located inside the sentences rather than primarily at the beginning. Nonetheless, they link parts of whole thoughts. By making it easy for the reader to follow the pattern of your thought, you maintain flow and help keep the reader focused on what is important.

4.3.2 Maintain Focus

KEY IDEA: The topic sentence is the most powerful writing tool for helping the reader focus on critical information.

When it comes to helping the reader focus on critical information, the most powerful writing tool you have is the topic sentence. Think about how you read paragraphs. Usually you read the first sentence for a clue as to what the paragraph is about. (Every so often, a topic sentence will come second or third, after a sentence or two that link to a previous paragraph.) Then you want to hear mainly about *that* subject; if the paragraph tries to cover too many subjects or if it has no clear topic sentence, you will understandably become confused. The topic sentence is like a welcoming door to a house: If the door is closed and locked, you will feel unwelcome and not go in. If it is slightly open and you peek inside, you will want to know where you are entering. Is this a place of good, clear information?

Consider the following paragraph:

> The State does not appropriate funds to maintain and/or build surface parking lots, parking facilities, or parking garages. All parking on campus, including surface parking and the parking garage, is operated and maintained from the parking system revenues. Revenues include the annual parking permit fee, incomes from citations issued by The University Police, and fees charged to use the University parking garage. The construction of a second parking garage is necessary. To meet the required funding to construct this parking garage, it will be necessary to increase the campus parking revenue by raising the annual parking permit fee. The parking permit fee will increase over a two-year period to accomplish this.

Does the first sentence make you want to "enter" the paragraph? Do you feel that the first sentence is a good indicator of the subject matter covered by the rest of the paragraph? If you answered "no" to both questions, you are not alone. Most people would not read this paragraph all the way through. Knowing the sources of the state's funds does not seem important at the beginning because we are not given a clue as to the real subject matter of this paragraph. Of course, the real subject is bad news: The parking fee will go up. So we can understand why the writer did not want to use the last sentence—*The parking permit fee will increase over a two-year period*—as the topic sentence. But the problem is that hiding bad news may mean that someone simply does not read the entire paragraph and is therefore shocked when receiving a higher bill in the mail. If you really do not want to inform your readers, do not write anything, because badly organized writing will only *misinform* them. If you *do* want to inform them, figure out a way of organizing the paragraph so that you keep the reader focused on the most positive message—in this case, the fact that a garage will be built to help solve the university's parking problems.

Practice Question: Topic Sentences

Looking at the foregoing paragraph about constructing a new parking garage at a university, can you come up with a topic sentence that emphasizes a positive message? What would be a better second sentence?

Suggested Answer

- Here is one suggestion for a better topic sentence:
  To solve the problem of increased traffic on campus, the university is building a second parking garage.

This sentence focuses on the solution to the problem of increased numbers of cars on campus.

- Here is a suggestion for the second sentence:
  To fund this initiative, the parking fee will increase over a two-year period.

Another way to maintain focus in every sentence of a paragraph is to keep the reader's eye on the ball—that is, on the most important pieces of information. Consider this pair of sentences:

> The proposal does not conform to our research goals as outlined in the five-year strategic plan. For instance, the potential for discovering new laser techniques is discussed only briefly and does not form a major part of the proposal.

The topic sentence here seems fine. But the second sentence does not tie in well with the first, in spite of the helpful transitional phrase "For instance." The second sentence is confusing because it does not emphasize at the beginning what is important: the proposal. In the following revision, the noun that is the main subject is repeated in the position of subject:

> The **proposal** does not conform to our research goals as outlined in the five-year strategic plan. For instance, the **proposal** fails to emphasize our potential for discovering new laser techniques; these techniques are discussed only briefly.

Placing "the proposal" in the position of subject of the second sentence greatly increases our comprehension of the content. We know now to keep focusing on the proposal and what it does and does not do.

KEY IDEA

> **To Maintain Focus:**
>
> - Tie new information to old information.
> - Place repeated information in the subject position.

Research has shown clearly that readers want new information to be tied explicitly to information they already know. Study this passage from *Physics Today* (August 2002, p. 21) and notice how seamlessly it weaves new information into the fabric of given information by repeating key words and concepts in the "subject" position:

> After setting a **wave** in motion, the researchers take an **image** of the tray with a video camera. **The wave** is brighter than its unexcited surroundings, and its size and location are easily **measured**. **These measurements** go into the feedback step: determining the pointing direction and brightness of the reaction-controlling spotlight. By comparing the current **image** with an image taken two seconds earlier, the researchers can tell whether the **wave** needs a bigger or smaller dose of light to maintain its stability.

This passage also makes good use of parallelism to channel the reader's attention. What is parallelism? Read on ....

4.3.3 Maintain Parallelism

When you present a series of facts, findings, or steps in a process, you want readers to remember how the items are similar and dissimilar. Mostly, you want them to remember the facts or steps themselves. To make clear the similarity of certain items and to aid reader's memories, you need to use parallel structure (parallelism). A sentence such as this one is unnecessarily wordy and potentially confusing because it does not maintain parallelism:

> The architectural engineer finished the mechanical plan, made five copies, and then it was forwarded to the client.

Of the three actions the engineer took, two are presented in the active voice and one in the passive voice. Maintaining parallelism means being consistent with your grammatical constructions for each item in your list, especially with your verb forms.

To maintain parallelism, you must choose a grammatical form of presentation (based on parts of speech, usually) and stick to it. Here is a set of instructions that is not in parallel form:

Recommended Procedure for Characterizing Emodin:

1. *Obtain emodin from reliable chemical manufacturing company. (See list.)*
2. *Identification should be made with nuclear magnetic resonance spectroscopy.*
3. *Determining purity should be done by elemental analyses.*

In this example, if we agree that the first instruction ("Obtain emodin . . .") is the best form for a set of instructions, then we should stick to that form for the other items. Here, the chosen form sets up each item in the following sequence of parts of speech:

command form of a **verb/ noun** *as object/* **prepositional phrase**
obtain/ emodin/ from ... company

If we rewrite steps 2 and 3 in parallel form with step 1, we come up with something like this revision:

Recommended Procedure for Characterizing Emodin:

1. ***Obtain*** *emodin from reliable chemical manufacturing company. (See list.)*
2. ***Identify*** *emodin by using nuclear magnetic resonance spectroscopy.*
3. ***Determine*** *purity through elemental analyses.*

Present parallel ideas in parallel form in the following structures:

- lists;
- sentences (series or pairs).

Pay special attention to the first few words of each clause or phrase.
Remember that parallel constructions channel the reader's attention and aid his or her memory.

KEY IDEA

As we have seen, parallelism clarifies sentences as well as vertical lists. Let us return to our original example:

> The architectural engineer finished the mechanical plan, made five copies, and then **it was forwarded** to the client.

This sentence is easy to fix, right? We simply stick with the past-tense verb form of the first two actions. The final phrase then becomes active instead of passive: "...and then **forwarded it** to the client."

Here is an example that includes more than one "and":

> We researched and compared the performance capabilities and FAA records of each aircraft, **as well as conducting** a pilot survey.

The phrase after the comma is not in the same grammatical form as the other items in this sentence list, right? You have to change "conducting" to a past-tense verb: "conducted." Then you can make explicit the fact that the research involved at least two distinct steps—researching and comparing, and conducting a survey— not one big simultaneous step. Readers like processes and information to be explicitly broken down into understandable steps or ideas. Here is one possible revision:

> We researched and compared the performance capabilities and FAA records of each aircraft, and then **we conducted** a pilot survey.

4.4 GRAMMAR AND SENTENCE STRUCTURE: DIAGNOSING AND CURING THE PROBLEMS

Some problems with sentences are matters of clarity and style, and some problems are matters of grammatical correctness. Focus on the clarity first; then fix the grammar. There are several well-described methods available for pinpointing the weakness and wordiness of so many of our sentences. The best writing guides include Strunk and White's *The Elements of Style* (1979) and Richard Lanham's "Paramedic Method," developed in *Revising Business Prose* (2000). The best methods all work to counteract the vague and bureaucratic writing that Lanham calls the "Official Style." You may think the Official Style is the best way to write, because you see so much business and professional writing that is artificially pumped up with extra words and roundabout explanations. Actually, much of that writing is simply imitating a pompous style to which *those* writers were exposed. The fact is that most technical writers have never been taught how to revise their writing.

Let us look at a sentence on the experimental results of a study of liquid-jet-gas pumps:

> An important conclusion extracted from the result **is** that **there is** a correlation between the mixing length and the volumetric flow.

KEY IDEA: Stop using the constructions "there is" and "there are." State *what* is, not *that* it is.

Thinking about the words "An important conclusion ... is that there is ..." we realize how empty they are. Conclusions should simply be stated, not backed into. If we follow Richard Lanham's advice, we should improve our sentences by always asking the question, "Where is the action?" Most sentences focus on their verbs because we are most interested in what happened (or will happen or could happen). That something simply "is" is rarely enough information about it. The main action of the original sentence might be "extracted." But is the fact that the conclusion was "extracted" really important to the reader? Are we not interested mainly in the conclusion itself, in what the result demonstrates? Something important has been proven, or at least demonstrated. Can we express that result in a straightforward verb? Consider the following revision:

> The result demonstrates a correlation between mixing length and volumetric flow.

Pruning this particular sentence has also involved changing a passive construction ("conclusion extracted from... is") to an active construction ("result demonstrates").

We already know that excessive and inappropriate use of passive voice can be a problem for the reader, but how do you recognize passive voice in your own writing and test whether it is appropriate or not? Let us look again at a paragraph from the Work Remaining section of a progress report:

> The evaluation of diamond film based on the final criterion **will be completed** when the analysis of the qualities **is received** from the Institute for Advanced Technology (IAT). A request for the resource files on diamond film **will be submitted** to Bolton Manufacturing. An additional comparison of the results received from IAT **will be made** with information provided by Bolton. The evaluation of diamond dust **will be used** for the comparison of the other nanocrystalline coatings.

The passive-voice constructions are printed in boldface. To test for passive voice, just check each verb and see whether you can easily ask the question, "By whom?" For instance, "will be completed" brings up the question, "By whom?" If you have read the progress report up until this point, you have a pretty good idea that the agent of the action of "completing" is the writer, so you are probably not confused at that point in the paragraph. But you are forced to keep asking yourself, "By whom?" for every single action in this paragraph. There are, however, other potential agents of action (IAT and Bolton), so eventually you get tired of having to answer that darned question at least once in every sentence. Excessive use of passive voice wears us out! Reading is time consuming enough without having to worry about who is the agent of the action all the time!

Of course, not all uses of passive voice are wrong. Especially in technical writing, we may not need to know who is the doer of the action. Sometimes the sentence itself will answer the question, as in this example:

> The report **was disseminated** by the Department of Justice.

In that case, the verb is still passive, but the question gets answered, and so we are not confused. But if many sentences take this form, the document will be longer than it needs to be. As a writer, think about what is more important: the object of the action or the doer of the action.

Of course, putting verbs in active voice means inserting what is missing: the agent of the action. If the agent is the writer, technical style has long dictated that the writer may not appear as "I." In many contemporary journals, however, "we" is now perfectly acceptable. If you are told that you cannot use "I" or "we," then you

KEY IDEA: Excessive use of passive voice wears a reader out.

may want to restructure some of your sentences to include active verbs nonetheless. Consider the following example:

> The evaluation of diamond film based on the final criterion will be completed when the analysis of its qualities is received from the Institute for Advanced Technology (IAT). A request for the resource files on diamond film will be submitted to Bolton Manufacturing. An additional comparison of the results received from IAT will be made with information provided by Bolton. The evaluation of diamond dust will be used for the comparison of the other nanocrystalline coatings.

We may revise this paragraph as follows:

> The evaluation of diamond film based on the final criterion **will be completed** when the analysis of its qualities **is received** from the Institute for Advanced Technology (IAT). Once Bolton Manufacturing **responds** to a request for resource files on diamond film, an additional comparison of the results received from IAT **will be made**. The evaluation of diamond dust **will be used** for comparison with the other nanocrystalline coatings.

Of the five verbs with a subject in the foregoing paragraph, four are still in passive voice, but one is now in active voice—"responds." Turning just one verb form into active voice helps the flow of this paragraph and helps prune away other unnecessary words. Of course in this case—an industry progress report—there is no reason not to use "we." The writer is probably doing work for a client.

Another test for good writing is determining the number of prepositional phrases. Check out the original version of the last two sentences again:

> An additional comparison <u>of</u> the results received <u>from</u> IAT **will be made** <u>with</u> information provided <u>by</u> Bolton. The evaluation <u>of</u> diamond dust **will be used** <u>for</u> the comparison <u>of</u> the other nanocrystalline coatings.

In these two sentences, totaling only 33 words, there are seven almost-consecutive prepositional phrases. Those phrases slow down our reading, because prepositions do not involve action. They do establish relationship; something belongs *with* something else or comes *after* it. But too many in a row pull us away from the main thread of action that we are always trying to follow. As a rule, stick to no more than three prepositional phrases in a row. If we look again at our revised sentences, we can make them even stronger if we get rid of a couple of those phrases:

> Once Bolton Manufacturing sends us its resource files <u>on</u> diamond film, an additional comparison <u>with</u> the results received <u>from</u> IAT will be made. The evaluation <u>of</u> diamond dust will then provide a comparison <u>with</u> the other nanocrystalline coatings.

KEY IDEA

> **Remember: No more than three prepositional phrases in a row!**
>
> This report describes *the process <u>of</u> collecting data <u>on</u> incidents <u>of</u> outdoor burning statewide.*

4.5 WORDS: PICKY, PICKY

Paying attention to every word you use makes for good writing. You want to be sure that you are saying exactly what you mean. Be sure to keep a list handy of the words you know you tend to use improperly or are unsure of. See Table 4.3 for good examples of words that trip up many people.

Table 4.3 Words That are Often Misspelled or Misused

Diction: Watch out for these pairs of words!	
affect/effect	"Affect" is usually the verb.
	"Effect" is usually the noun.
among/between	"Among" for more than two.
cite/site	"Cite" is the indicates verb form of "citation".
	"Site" is the place.
criteria/criterion	"Criteria" is plural.
data/datum	"Data" is plural.
fewer/less	"Fewer" refers to number: *fewer tons*.
	"Less" refers to volume: *less capacity*
imply/infer	"Imply" is your intention.
	"Infer" is your discovery.
think/feel	Engineers generally "think" or "conclude" that something is the case; they are not paid to consult their feelings.

4.5.1 Verb Tenses

Many technical writers get confused about which verb tenses to use in which sections of a report. Do we say, "research show<u>s</u>" or "research show<u>ed</u>"? Proposals are easier in this regard because they describe a plan for doing something in the future: "We will do x, y, and z." Here are a few sensible assumptions about verb tense and when to use which.

Past tense is used mainly to describe the work you or others did to come up with the findings you have. Present tense is used in technical communication to indicate that you are speaking of something that is (apparently) eternally true— $e = mc^2$—or that you are speaking of what the reader is reading: "This report analyzes lab-test results for the air samples." Future tense is used solely for statements that project into the future with plans or visions. Do not refer to sections of your report in the past or the future tense. For instance, in the methodology section do not write, "The introduction described...," but rather, "The introduction describes...." Figure 4.1 gives more examples of how to use past, present, and future tense in specific sections of a report or paper.

Past Tense:

- **describes what you did**
- **speaks biographically of another researcher's actions**
- **sets up a historical continuum**

Use mainly in Procedures, Methodology, and Background sections (also in parts of the Introduction).

Example:

"Researchers **have** long **known** that microbes can destroy contaminants in soil." (historical continuum)

Present Tense:

- **states theory or established knowledge**
- **says what your document does**

Use mainly in Introduction and Conclusions.

Example:

"This research shows that microbes **can destroy** contaminants in soil." (conclusion)

Future Tense:

- **outlines recommendations for future work or action**
- **envisions a future state of affairs**

Use mainly in the Recommendations section.

"Implementing this water treatment system will ensure safe drinking water.

Do not use this tense to state what your report does!

"This report ~~will~~ present the results of the study."

"This report presents the results of the study."

Figure 4.1
Uses of past, present, and future tense in engineering reports and papers.

4.6 PUNCTUATION: WHY SHOULD I CARE?

Do you think of punctuation as a set of unfathomable rules that mostly do not matter, except that teachers like to make corrections on your papers? If so, you are not alone. Many students get conflicting messages about various marks of punctuation, especially commas. Should the comma always follow an introductory clause, as in the following sentence?

> After the initial tests were run, the material was sealed in an airtight container.

Should the comma always follow the last item in a series, coming before the final "and" or "or," as in the following sentence?

> The three steps of this process are gas separation, sulfur recovery, and excess gas disposal.

KEY IDEA: Punctuation helps clarify your writing, but punctuation guidelines do vary among disciplines.

Actually, neither of these questions has one right answer; as a rule of grammar, the comma is generally considered to be optional in both cases. Practicing writers, however, will tell you that using the comma in these two places often helps reader comprehension and provides fewer confusing choices for the writer. In particular, the final comma in a series sometimes has to be there, such as if the items themselves in the series contain "and." See the following example in which the final comma is necessary to avoid confusion:

> Engineers learned to isolate **and** separate the gasses, identify **and** recover the sulfur, **and** dispose of the excess gas.

Using the final comma routinely means that the writer doesn't have to worry about whether it is required or not. At least one revered style guide, however, advises *not* using the comma when the only "and" is the final "and" (Strunk and White, 1979).

The main reason to bother about punctuation is to control and clarify your meaning in any given sentence. Even hyphens can be critical in conveying the correct meaning. For instance, which of these phrases is correct?

fast-sailing ship
fast sailing ship

Well, of course, it depends on whether you are referring to a ship (any kind) that sails fast or a sailing ship (in particular) that is fast. Many engineering disciplines have noun phrases that are commonly used without hyphens. Remember to think of your audience before you omit the hyphens from a phrase such as this: *antibiotic resistant infections*. If you want to be sure that no reader will mistakenly assume that you are talking about resistant infections *caused by* antibiotics, then you should write the phrase this way: *antibiotic-resistant infections*. Check with your instructor or boss about the punctuation of commonly used phrases in your field, and read around in your field's technical journals to see how the phrases are handled there.

In general, the best practice is to familiarize yourself with punctuation rules as outlined in some reference guide and then simply place the guide on your desk for easy reference when writing. Here is an annotated list of a few of the best guides:

Strunk, W., and White, E.B. 1979. *The Elements of Style*, **3d. ed. Macmillan: New York.**

Though originally written almost 50 years ago, this concise guide to writing style is a bible for many technical professionals. (Actually, the original book was privately printed by William Strunk in 1918!) It covers only those elements and issues of punctuation that are most relevant, according to the authors—issues such as whether to incorporate the comma in a series before the final "and." But because the book is short and written in the style it recommends—direct, brief, and precise—its "essential" rules of punctuation, grammar, and style are easy to remember and have stood many, many writers in good stead for decades.

Chicago Manual of Style, **15th ed., 2003. University of Chicago Press: Chicago.**

This is the mothership of style manuals. This big volume focuses on style and publication issues, but also has an extensive chapter on punctuation (almost 40 pages long) that is a terrific reference tool for engineers.

Harris, M. 2003. *Prentice Hall Reference Guide to Grammar and Usage*, **5th ed., Prentice Hall: Upper Saddle River, NJ.**

This volume is more reader friendly than the mammoth *Chicago Manual*; it uses tabs, section indices, and other formatting techniques to make finding information easy. The chapter on punctuation is comprehensive without being overwhelming and includes useful exercises.

Nagle, J.G. 1996. *Handbook for Preparing Engineering Documents: From Concept to Completion.* **IEEE Press: New York.**
This handbook contains a nice summary, in an easy-to-read chart form, of the major punctuation rules and usage.

SUMMARY

- The keys to effective writing are to prune away the deadwood (vague and useless words), control passive voice, and maintain focus by linking thoughts together clearly.
- Empty pronouns drive readers crazy; "it or "this" at the beginning of a sentence means nothing without a specific noun to modify. Combining sentences can be a good way to avoid empty pronouns.
- If you can write a good paragraph, you have got it made. Effective paragraphs have a topic sentence that acts like a miniature thesis, and all the sentences that follow fit together logically and derive from that topic sentence. The direction of thought *moves*.
- Transitional words and phrases connect ideas and link data together. Making connections is what writing technical information is all about. Otherwise you are just making lists.
- Use a good style guide to pinpoint and fix grammatical errors and weaknesses of style in your sentences.

PROBLEMS

4.1. In the following paragraph, change the passive constructions (italics) to active verb constructions. Passive constructions are discussed in Section 4.1.

The evaluation of diamond film based on the final criterion *will be completed* when the analysis of the qualities *is received* from the Institute for Advanced Technology (IAT). A request for the resource files on diamond film *will be submitted* to Bolton Manufacturing. An additional comparison of the results received from IAT *will be made* with information provided by Bolton. The evaluation of diamond dust *will be used* for the comparison of the other nanocrystalline coatings.

4.2. After changing the passive verb constructions in the paragraph in Problem 4.1, answer these questions:

Does this exercise force you to add additional words?

Are those words mainly transitions, or do they have another function?

How do these additional words help keep the paragraph moving along?

4.3. Write a list of all the words or phrases that could fill in the blank in the following pair of sentences. Which transitional word or phrase is best and why?

The technician heated the material to a temperature of 100° C. _____, the material changed color slightly.

4.4. Rewrite the pair of sentences in Problem 4.3 as one sentence. How did you punctuate it? Is the information clearer when expressed as one sentence or as two sentences?

4.5. Add correct punctuation to the following sentence:

> **Spending limit balances will carry forward for some financial accounts however we will remove expenditure limits against IT accounts on August 31 2004.**

4.6. True or false? In a report, you use past tense to state theory or established knowledge.

4.7. True or false? The word "criteria" is used properly in the following sentence:

> **The design criteria is one reason we decided to shorten the length of the rod.**

4.8. Which word in the following sentence is used improperly?

> **Among the two of us researchers, there can be no disagreement about methodology.**

FURTHER READING

Ballstaedt, S-P and Mandl, H. 1987. "Influencing the Degree of Reading Comprehension," in *Knowledge Aided Information Processing*, Van Der Meer, E., and Hoffmann, J., eds. Elsevier Science Publishers: Amsterdam.

This collection of essays on recent theories of cognitive psychology is far too technical for most readers, but the cited article gives a good idea of the complex interplay of elements—memory, prior knowledge, semantic analysis, and levels of processing, just to name a few — that enable us to understand what we read.

Chicago Manual of Style, 15th ed., 2003. University of Chicago Press: Chicago.

This is the mother ship of style manuals. This big volume contains more than you'll ever want to know about manuscript preparation, documentation, and the publishing process, but it will also have definitive answers for you on all those niggling little questions about style (do you write out the number twelve or use the Arabic numeral?). The sections on "Grammar and Usage" and "Punctuation" are a terrific reference tool for engineers. Most discipline-specific style manuals in engineering are based on the *Chicago Manual*.

Day, C. 2002. "Remotely Sensed Neutrons and Gamma Rays Reveal Ice beneath the Martian Surface" and "Choreographing Wave Propagation in Excitable Media," *Physics Today*, August 2002, vol. 55, no. 8, pp. 16–17 and 21.

Reading science and engineering magazines intended for a mixed audience (technical and semitechnical) can provide models for good technical writing. *Physics Today* is one such magazine; each issue contains thoughtful articles on very technical subjects written for anyone with an interest in physics or in any of the fields deriving from physics (such as engineering). See next entry for another example of good technical writing.

Fitzgerald, R. 2000. "Phase-Sensitive X-Ray Imaging," *Physics Today*, July 2000, vol. 53, no.7, pp. 23–26.

Lanham, R. 2000. *Revising Business Prose*, 4th ed. Allyn & Bacon: Needham Heights, MA.

Richard Lanham's book contains a wonderful discussion of how to tighten your sentences and strengthen your meaning. He calls this procedure the "Paramedic Method" for diagnosing and curing wordy sentences. Read the whole book (it is short), but look especially at the first chapter, "Who's Kicking Who" (if you caught the error in grammar, you are right—Lanham wants to get your attention).

Strunk, W., and White, E.B. 1979. *The Elements of Style*, 3d. ed. Macmillan: New York.

This book is the last word for many technical professionals on style and grammar. It covers only those style guidelines the authors considered most important or most often abused, but the book's frank, direct, and directive style has delighted many people and helped them remember rules such as "Do not join independent clauses with a comma." The effect of this book has been extraordinary for almost a century. It was originally written (by William Strunk) in 1918!

Zinsser, W. K. 1990. *On Writing Well: An Informal Guide to Writing Nonfiction*, 25th anniversary ed. Harper Collins: New York.

William Zinsser is a prolific writer whose books and journalism pieces cover subjects from jazz musicians to the art of Roger Tory Peterson. This book on writing has become a classic for its crisp demonstration of the value of simplicity in writing and for its explanation of writing as a transaction between author and reader. Chapter 15 covers science writing and technical writing.

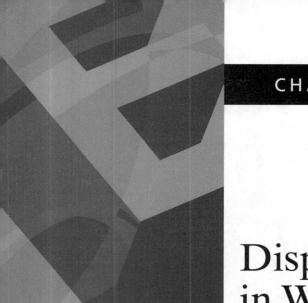

5

Displaying Data in Written Documents

Objectives

By reading this chapter, you will learn the following:

- the importance of displaying data;
- how to design and construct appropriate representations of your data for written documents;
- how to integrate graphical representations into your documents;
- how to explain in words the significance of each graphic;
- how to use mathematical and chemical notation.

5.1 INTRODUCTION

A picture is said to be worth a thousand words. And so is a table, a scatter plot, or a bar chart. Figures and tables are often the heart of an engineering document. They display the data, and the exact relationships among data, that will give life and meaning to all interpretation and conclusions. In addition, representing data graphically allows us to show relationships among the data quickly and effectively. For example, it would take far longer to try to explain in words all of the relationships expressed in Figure 5.1.

If you were to explain all parts of Figure 5.1, you would have to write something like this:

> *Air-quality modeling integrates emissions data (from multiple sources), meteorological information, and chemical mechanisms to determine pollutant production over time and space. Emissions are produced by natural organisms as well as by man-made sources such as automobiles. The pollutants produced are many, and the current work is concerned with how these pollutants interact to produce ozone in concentrations greater than current standards deemed safe for human health. This work is concerned primarily with ozone concentrations in the northeastern U.S.*

However, you cannot make the mistake of thinking that a good technical communicator does not need words to describe the significance of these relationships. A picture may be worth a thousand words, but it does not necessarily replace all words. Some people are not used to reading figures and tables, and they need to be oriented to what the figure shows. Other people know how to read figures, but they are reading quickly and could thereby miss the significance of the information revealed by the graphic. Using figures or tables *and* words allows you to communicate very effectively. By combining modes of delivery, you allow readers to use parallel processing in understanding the importance of your information: They can use both their visual and their verbal understanding. Especially when dealing with complex data, you need the

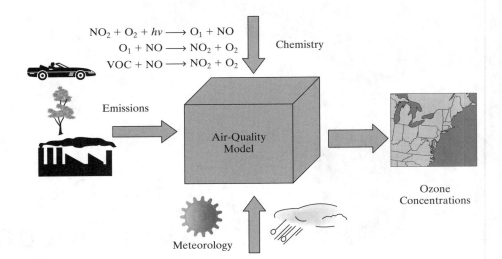

$$NO_2 + O_2 + hv \longrightarrow O_1 + NO$$
$$O_1 + NO \longrightarrow NO_2 + O_2$$
$$VOC + NO \longrightarrow NO_2 + O_2$$

Chemistry

Emissions

Air-Quality
Model

Ozone
Concentrations

Meteorology

Figure 5.1
A model for ozone concentrations in the northeastern U.S.

double power of words plus graphics to ensure that readers and listeners will get the total picture. Some people are more visual in their orientation toward learning, and some people are more verbal. Most engineers tend to be visual, but your readers may not be engineers.

Why use graphics in technical documents?

- to show complex data in a simplified form
- to show a lot of data in one place
- to emphasize relationships
- to help the reader remember
- to allow parallel processing of information

Graphics are also a very important part of presenting information verbally. It is difficult to imagine a technical presentation that does not exhibit well-projected, readable figures. The design rules governing presentation graphics, however, are somewhat different from those for graphics in written documents. This chapter deals with how to design and use graphics in *documents*.

5.2 DISPLAYING YOUR DATA

As an engineer, you will collect data in various ways. Sometimes you will model a system or process, providing certain inputs and extracting outputs. Sometimes you will use mathematical calculations to test a hypothesis or prove a theory. And very commonly, you will use experiments, in the lab or in the field, to yield information that helps solve problems. Let us posit a situation in which you work for a manufacturing company as an air-pollution control expert. Your job is to find economical ways of cleaning the gas waste stream produced by the manufacturing process. Biofiltration is a promising technique that uses microorganisms to reduce pollutants in the gas stream, but you need to find out how large a biofiltration system would be needed at the plant and how much it would cost. A typical system uses a packed bed of biological material inside a column through which the gas stream is run for treatment. Figure 5.2 is a schematic identifying key components of the system.

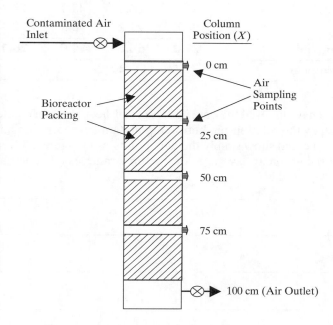

Figure 5.2
Schematic of the experimental bioreactor.

You set about designing and conducting a series of experiments to show the relationship between the size of the system and the amount of pollutants in the gas stream that is destroyed by the microorganisms. You run the experiments, collect the data by taking samples of the gas at various positions within the column, and record the data. At that point, your communication work begins, because you must present the data to various managers within your company and help them come to a decision about what size system to use. You must first display the data for your own benefit, so you can determine at what position in the column enough pollutants are destroyed that the column has done its work and is no longer necessary; a shorter column makes for a cheaper treatment system. Then you must show these data and this relationship (between position in the column and concentration of pollutants) to others in a comprehensible, visual form. But how?

Well, to understand the data yourself, you will use a spreadsheet program such as Microsoft® Excel. You will enter the sampling data (concentration of pollutants) that you recorded, with three samples for each position in the column. Let us say you recorded data in triplicate at the inlet point, at the outlet point, and at 25, 50, and 75 cm from the inlet of the column. Since you will want to take the average of the three readings for each sample, your data sheet would look like Table 5.1 (where pollutant concentration is measured in milligrams per cubic meter).

KEY IDEA: You will create different displays of your data for different audiences, and for yourself.

Table 5.1 Excel Worksheet Containing Sample Data for an Experimental Biofiltration System

Position (cm)	0	25	50	75	100
Concentration (mg/m³)					
sample 1	101.0	68.0	45.5	28.8	15.0
sample 2	98.4	66.5	43.0	27.3	13.1
sample 3	97.7	71.2	46.3	22.3	16.8
Average (mg/m³)	**99.0**	**68.6**	**44.9**	**26.1**	**15.0**
Standard Deviation	1.74	2.40	1.72	3.40	1.85

Once you have entered the data in the spreadsheet, you can create many kinds of graphs to show the relationship between pollutant concentration and position in the column. You could show simply the data points, as in Figure 5.3 (remember that the points on the graph are averages of the three samples taken at each position).

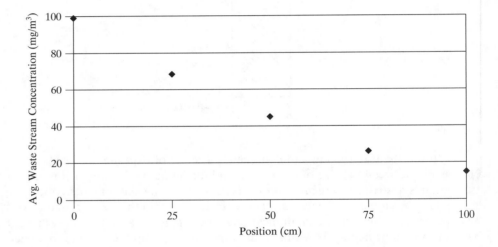

Figure 5.3
Relationship of position in the biofiltration column and pollutant concentration of the gas waste stream [data points only].

If you want to show more clearly the *trend* in the data, you could connect the data points (as in Figure 5.4a) or fit the data by using a regression model (as in Figure 5.4b). Excel contains a tool called "trendline" that provides various regression models. Be sure to indicate the equation of the trendline on the figure.

If you wanted to show more clearly the *scatter* in your data (emphasizing the fact that not every one of the three samples taken at the same position had exactly the same concentration), you could use error bars, which plot the standard deviation of the data at each point, as in Figure 5.5.

So, now you would be ready to put together a report for a variety of audiences, from the other project engineers to the head of your division, and even the chief financial officer. Of course, your presentation or report should begin with a schematic of the system, showing the column and where the gas stream enters and exits. Remember, we began this discussion about displaying data with a drawing of the treatment column. Be sure to use visuals to orient your reader to as many aspects of your experimental conditions as possible.

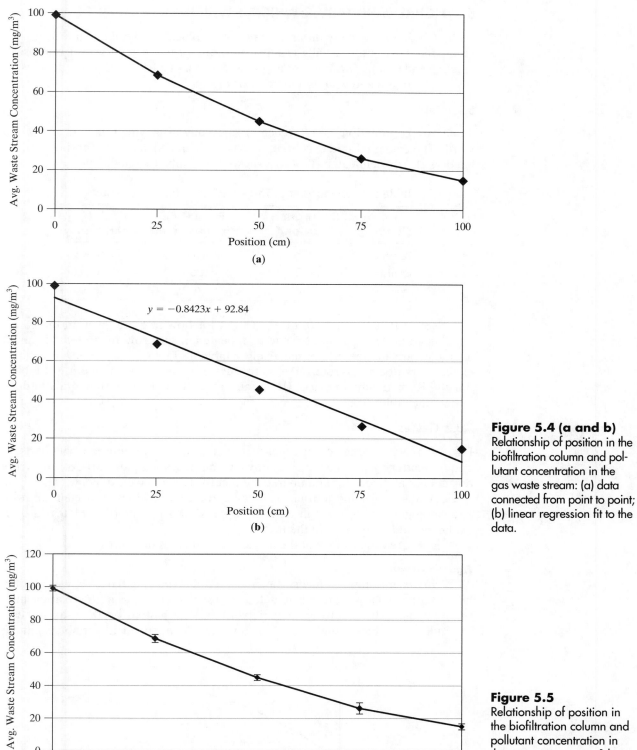

Figure 5.4 (a and b)
Relationship of position in the biofiltration column and pollutant concentration in the gas waste stream: (a) data connected from point to point; (b) linear regression fit to the data.

Figure 5.5
Relationship of position in the biofiltration column and pollutant concentration in the gas waste stream [showing standard deviation of data at each point].

5.3 FIGURES AND TABLES: WHICH IS WHICH?

Although there are many different types of graphical display, in engineering documents these types are usually grouped into one of two categories for purposes of labeling and numbering: figures and tables. So, first of all, let us answer the question, "What is a figure and what is a table, and why?"

5.3.1 Tables

Tables are compilations of reference data, usually displaying numbers and/or keywords. The data are listed in columns (vertical elements) and rows (horizontal), with headers clearly delineated. Table 5.2 shows a typically designed table.

Table 5.2 Comparison of Three Air Filters for an Aquarium

Filter	Calibration Factor	Flow Rate (gallons/hour)	Efficiency
Power	4–5	1000	Least
Canister	1–2	250	Moderate
Wet/Dry	3–4	700	Highest

Notice that the units of measurement for flow rate are given in the column heading, so that they do not have to be repeated in each row. In written documents, all tables are numbered consecutively—Table 1, Table 2, etc.—and each has an explanatory title and is referred to in the text. Sometimes the numbering of tables is done in Roman numerals (e.g., III, IV, and V) instead of Arabic numerals, to differentiate the tables from figures.

5.3.2 Figures

Figures are what we call every other sort of graphical representation: bar charts, line graphs, scatter plots, schematics, plan views, photographs, pie charts, etc. In written documents, all figures are numbered consecutively and have explanatory labels and titles. A few books and journals still call certain illustrations (especially drawings and photographs) "plates" instead of "figures." All figures (and plates) are also explained and referred to in the text.

Here are descriptions of the more common types of figures.

Figures: Graphs

X–Y graphs are most engineers' favorite way of displaying data. They are particularly useful for showing trends in data over time and changes in the relationship of two variables. The independent variable is usually plotted on the *x*-axis and the dependent variable on the *y*-axis. In Figure 5.6, the independent variable is time (in

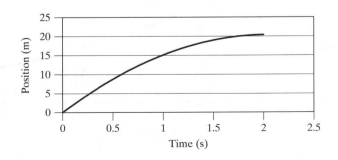

Figure 5.6
Distance a ball travels when tossed into the air.

seconds), and the dependent variable is position, measured in meters from the starting point (0). The graph shows how far a ball travels when it is tossed into the air until it starts to fall. In this graph, the slope of the curve mimics the physical reality that the graph expresses; however, graphs do not usually try to look like what they are depicting.

In Figure 5.7, the slope of the curve actually looks like the opposite of what is happening. This figure displays the velocity of a ball as it travels upward through the air to the point where it would start to fall and actually show a negative velocity.

A portion of the data set for Figures 5.6 and 5.7 is shown in Table 5.3. To graph the *distance* the ball travels, you would plot the values for the independent variable in column 1 (time) and the dependent variable in column 2 (position). To graph the *velocity* of the ball over time, you would plot columns 1 (time) and 3 (velocity). These data can be generated by the formulas you enter for velocity and by the ranges you specify for time and position. Since acceleration is a constant governed by gravity (*g*), you probably would not want to graph it, but you should include it in your explanation of the figure and, whenever possible, show such relevant data on the figure itself.

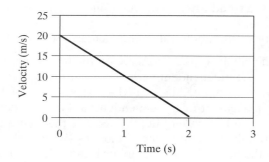

Figure 5.7
Velocity of ball tossed into the air (at constant acceleration of -9.8 meters/second^2).

Table 5.3 Data Set for Figures 5.6 and 5.7

INITIAL POSITION	0		
initial velocity	20		
time (s)	position (m)	velocity (m/s)	acceleration (m/s^2)
0	0	20	−9.8
0.5	8.775	15.1	−9.8
1	15.1	10.2	−9.8
1.5	18.975	5.3	−9.8
2	20.4	0.4	−9.8

Line graphs are useful to show changes in one variable over time or differences in values for certain items. You use line graphs when the *x*-axis is showing labels or increments, not numerical values. Thus, for example, you could show the differences in grades among four students as displayed in Figure 5.8.

Most line graphs could also be constructed as bar charts. Do not confuse line graphs with scatter plots that have a line drawn through the data points. Engineers do not use line graphs as often as *x*–*y* scatter plots because the former do not plot the relationship between two variables. For line graphs to make sense, the increments along the *x*-axis must be equal.

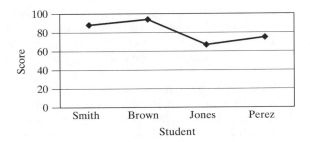

Figure 5.8
Final scores for four students.

Scatter plots show a relationship between variables for a particular phenomenon or quality at precise points. For example, Figure 5.9 shows measurements of soil samples from an offshore boring that was drilled from an oil production platform. The measurements are of the undrained shear strength of the soil, which is an indication of how strong the soil is. In the graph, the strength is shown in relation to the depth (below the mud line) of the sample taken. The important conclusions from this plot are twofold:

1. The plot shows a trend wherein the undrained shear strength tends to increase with depth.
2. The amount of scatter about this trend increases with increasing depth.

Notice that the labels for the *x*-axis are on the top of the figure rather than at the bottom and that the independent variable (depth) is plotted along the ordinate (the *y*-axis) instead of along the abscissa (the *x*-axis) as is usual. This exception to the rule governing placement of the dependent and independent variables occurs most commonly for graphs that plot depth. Because the concept of depth is inherently visual, we normally plot depth where it makes visual sense—along the vertical axis. Note also that the number *n* of data points is indicated on the figure. Whenever possible, place such helpful information right on the figure itself. If the information is more complex, it may be more appropriate to put it in a caption. If not, you will want to discuss it in your text, along with any other information about the experimental, modeling, or mathematical procedure behind the graphic.

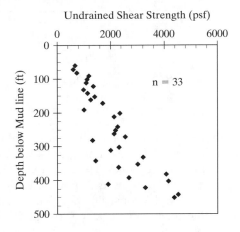

Figure 5.9
Relationship of undrained shear strength and depth for soil samples taken in the Gulf of Mexico, July 2001.
Graph courtesy of Dr. Robert Gilbert, University of Texas at Austin.

Figures: Charts

Bar charts are probably the most widely used graphic in the popular press, if not in engineering reports, because they are easy to understand and multifunctional. They come in many varieties, too, including histograms and Gantt charts. Comparing the sizes of bars is an easy way to compare differences in values or amounts of particular items. The prevalence of bar charts in annual reports suggests that cost comparisons are a natural subject for these kinds of charts. Figure 5.10 compares the costs of various air purifiers.

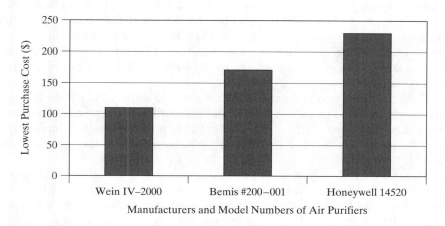

Figure 5.10
Lowest purchase costs found for Bemis, Honeywell, and Wein air purifiers.

Area charts, or stacked charts, are really a series of line graphs piled on top of each other. They are tricky to read because, beyond the bottom line, you do not read from the beginning of the scale: You separate out the "chunk" that is bounded by two lines and figure out the values within that chunk. Most nontechnical people, and even a lot of engineers, tend to read stacked charts improperly. Consequently, this type of chart should be used only rarely in technical documents. Consider Figure 5.11 and ask yourself how to determine the number of men who got degrees from this university in 1984–1985. In order to figure out the number of men receiving degrees in 1984–1985, you have to estimate how many are indicated in the chunk delineated by the arrows. Such an estimate is not very precise, and many people would be tempted to estimate the number by starting at the very bottom of the graph instead of at the line at the top of the "Women" chunk, adding about 100 to the correct number. The correct number is 1,240.

Pie charts are often used to depict constituents of a whole. These kinds of charts are seen everywhere, especially in magazines and newspapers, because they

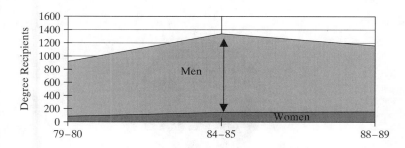

Figure 5.11
University of Malibu degree recipients by gender [hard to read].

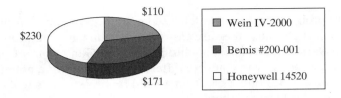

Figure 5.12
Lowest purchase costs of three air purifiers.

are ostensibly easy to read. They *are* easy to read, but only if they are designed and used well. Figure 5.12 displays the cost breakdown for the air purifiers (discussed previously) as a pie chart. Designing this chart in three dimensions doesn't make much sense, however, and lends a bit of distortion to the figure. The slices facing toward us seem proportionately larger than they are because they (seemingly) show a third dimension.

Pie charts really come into their own when you display the values as percentages of a whole, as in Figure 5.13. The reader can easily see the pieces as representative of the portion each material contributes to the total construction budget. In this figure, the pieces are "exploded," which emphasizes their relative size (and therefore value) even more.

Be sure that when you use a pie chart, your goal is to focus on the whole of something: a whole year, a whole graduating class, etc. In the pie chart in Figure 5.14, the importance of the whole—the total number of graduates from this university in the years 1990–1996 is not immediately clear. Why is that number important, especially when it is not even given? It might be important for some analysis of recent alumni, for example, *but the graphic does not clarify this information.*

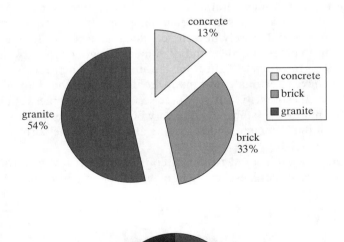

Figure 5.13
Relative contribution of exterior materials to construction budget.

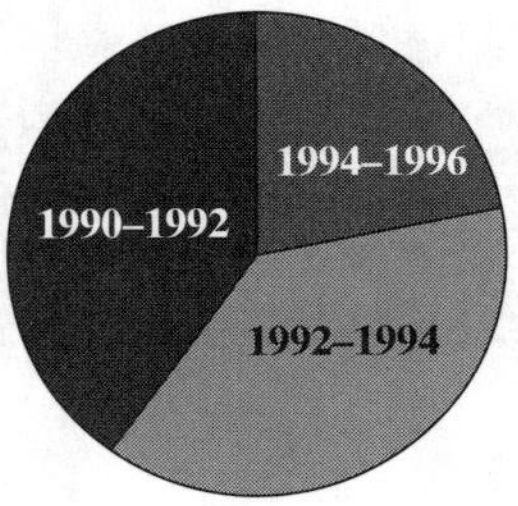

Figure 5.14
University of Malibu degree recipients, 1990–1996 [inappropriate pie chart].

Flowcharts make visual a process that otherwise would be more difficult to conceptualize and understand. They can depict processes as simple as activities in a workday, or they can depict more complicated processes, rendering them simpler to understand. Figure 5.15 shows a flowchart depicting the performance-review process in a company.

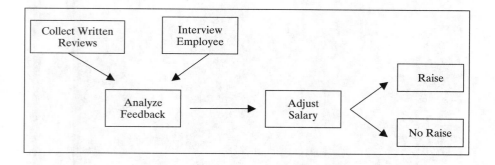

Figure 5.15
Performance-review process for Integrated Documentation, Inc.

Figures: Drawings and Schematics

Drawings and schematics simplify reality and thereby show the essential elements necessary to understand a particular problem or phenomenon. Figure 5.16 is a schematic of a bioreactor. The actual column, made from stainless steel, does not look exactly like the schematic in Figure 5.16, but the drawing shows all that is essential to understand how the experimental work was accomplished using this bioreactor. We see that the contaminated air stream was fed into the column at the top, that samples of air were taken every 25 centimeters, and that the column is 100 cm long. The figure also establishes that, for purposes of graphs showing results of the sampling, "0" is the top of the column, where the air is first fed in.

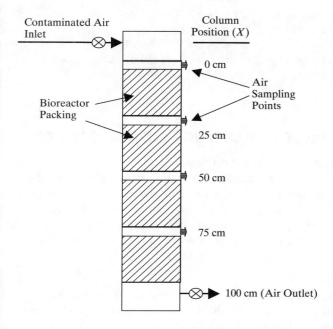

Figure 5.16
Schematic of an experimental bioreactor.

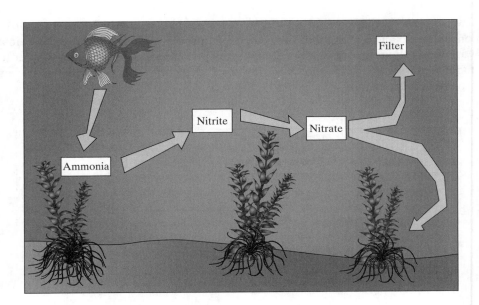

Figure 5.17
Process of pollution in an aquarium.

Figures: Combinations

Some figures artfully combine the attributes of several types of figures, as in Figure 5.17, a drawing and flowchart depicting the process of pollution in an aquarium. Be careful of trying to combine too many types of figures and thereby failing to help the reader understand information quickly. Figure 5.17 works because of its elegant simplicity.

Guidelines for Graphics: Engineering Conventions

- In graphs, use lines to illustrate predictions of models, and points to indicate data values found through experiment, especially when comparing the two types of information.
- Follow the convention for direction: When tables or figures are printed sideways on a piece of paper, the paper should be turned clockwise for the table or figure to be read)

5.4 GUIDELINES FOR DESIGNING FIGURES AND TABLES

Because figures and tables are often the heart of an engineering document, your figures must be clear and correct. If they are not, your methodology begins to seem questionable to the reader, and the foundation for your interpretive judgments crumbles. A picture may be worth a thousand words, but a graphic will be worth less than nothing if it represents the data inaccurately or confusingly. How do you construct useful and readable figures and tables?

5.4.1 Graphs

Here are some instructions for constructing a clear, readable graph. The abscissa is the horizontal axis, which typically displays values for the independent variable. The ordinate is the vertical axis, which typically displays values for the dependent variable. The point of origin of both the ordinate and the abscissa is generally zero, although there is a lot of disagreement about the importance of adhering to this guideline. In displaying certain data, zero may not be a meaningful value, and having to begin the ordinate with zero may mean creating a scale that does not capture the relevant data. To ensure that your audience will not misunderstand the graph for such a case (especially when you are unsure about the audience's technical background), consider starting the ordinate values at zero and then inserting a zig-zag break mark to indicate that the scale is being broken for the sake of displaying the relevant data. In Figure 5.18, all of the relevant values for the dependent variable (the STC values) were between 50 and 58 decibels, so the author started the *y*-axis at zero but then jumped to 50 to begin the next interval. The break mark indicates this jump. Unfortunately, Excel does not contain a feature that will insert a break mark. Other programs, such as SigmaGraph, do contain such a feature.

Both axes should be properly labeled. These labels should use words and symbols whenever possible and include the units of measurement; using both words and symbols can tie the figure well to the text. Common abbreviations may be used; when in doubt, however, spell out the word. In Figure 5.18, dB is the common abbreviation for "decibels."

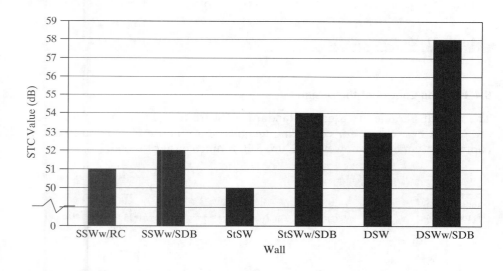

Figure 5.18
Performance measured as Sound Transmission Class (STC) values for each wall design. *Source:* Simmons, Leslie. *Construction: Principles, Materials, and Methods.* 2001.

5.4.2 Tables

Tables can be useful for showing comparisons of and changes in data, but they can easily become overcrowded and hard to read. Be sure to highlight important information in tables, such as totals or especially significant numbers or words. Color is effective as a highlighter, but remember that someone may make black-and-white copies of your table. For example, the original version of Table 5.4 had red numbers indicating unacceptable settlement. But in a black-and-white copy those numbers are lost in a sea of other black-and-white numbers, and thus the point of this table (that certain sizes of piles allow too much settlement of a structure) is not as clear. Also, the caption has to be rewritten, since red will not show up in the black-and-white copy. You can use shading to differentiate.

Tables often give a more compete "story" than a line graph (e.g., tables give specific numbers, while graphs typically present only approximate numbers), but generally they are not as effective as more pictorial forms. Consider including tables in appendices and creating line or bar graphs to summarize the same data in the body of your document.

Table 5.4 Comparison of Settling for Various Types of Piles

Piles that exceed 0.25 inches of settlement are shown in red. Piles that are shaded in gray were discarded. *[Author's note: Red cannot be seen in black-and-white reproduction.]*

Concrete Pile			Hollow Steel Pile			Concrete-Filled Pile		
Diam (in)	Len (ft)	Sett (in)	Diam (in)	Len (ft)	Sett (in)	Diam (in)	Len (ft)	Sett (in)
12	50	0.19	12	50	0.07	12	50	0.19
12	55	0.37	12	55	0.18	12	55	0.37
12	60	0.36	12	60	0.17	12	60	0.36
12	65	0.61	12	65	0.28	12	65	0.61
14	50	0.20	14	50	0.10	14	50	0.20
14	55	0.33	14	55	0.16	14	55	0.33
14	60	0.43	14	60	0.21	14	60	0.43
14	65	0.68	14	65	0.26	14	65	0.58
16	50	0.23	16	50	0.10	16	50	0.23
16	55	0.32	16	55	0.20	16	55	0.32
16	60	0.41	16	60	0.42	16	60	0.41
16	65	0.43	16	65	0.43	16	65	0.43

5.4.3 Taking Control of the Design

The default parameters used in standard software for generating figures are not always the best choice for conveying technical information. Figures 5.19 and 5.20 contain the same data; however, Figure 5.20 is much easier to read.

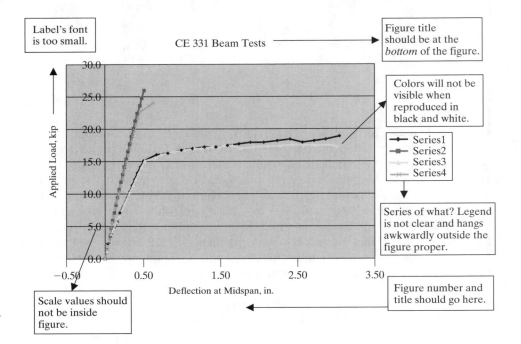

Figure 5.19
Concrete beam deflections at various loads [badly formatted graph].

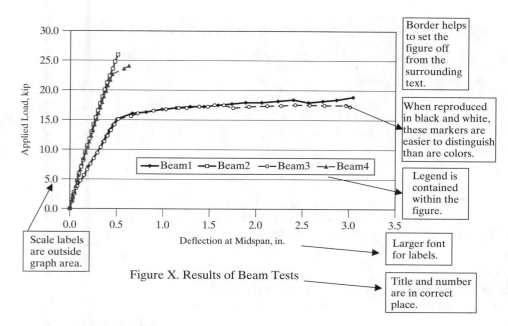

Figure X. Results of Beam Tests

Figure 5.20
Concrete beam deflections at various loads [well formatted].

The modified graph is easier to read because it uses a larger font and has a larger area in which to display the data. The modified graph also has a figure number and a meaningful title in the proper place.

Note that Figures 5.19 and 5.20 seem to flout the rule that the independent variable is plotted on the abscissa: Deflection, or strain, is dependent on the load (or stress) applied in the experiment. In structural engineering, however, stress–strain curves are always plotted such that strain is plotted horizontally and stress is plotted vertically. Be sure to become familiar with the conventions of your particular branch of engineering. A good way to obtain this familiarity is by reading around in the journals in your field. Notice how figures and tables are presented. Do the axes always cross at zero? Are units of measurement always abbreviated? Is the independent variable always plotted on the *x*-axis? Are tables numbered with a different style than are figures?

Consider the following items when preparing your figures:

- Gray backgrounds, such as that in Figure 5.19, look nice on the screen, but the technical information tends to be lost when the figure is plotted in black and white.
- Whenever possible, select scales so that the axes cross at (0,0). In all cases, position the axis labels outside the graph area.
- It is not always easy to distinguish among the default colors and symbols when graphs are printed in black and white. Selecting open and filled markers is one way to distinguish one data set from another.
- Default text sizes from graphing programs are typically too small when graphs are imported into a word processor.
- Default line widths from graphing programs are often too thin when graphs are imported into a word processor.
- The location of the legend may decrease the available area for displaying technical information.
- A figure should have a number and a meaningful title placed below, not above, the figure.

PONDER THIS

PRACTICE QUESTION: DESIGNING TABLES

What changes make Table 5.6 easier and faster to read than Table 5.5?

Suggested Answer

- Text and labels are centered.
- Unnecessary horizontal lines are removed.
- Column and row headings are in boldface and single spaced.

KEY IDEA: Keep every graphic as simple and uncluttered as the complexity of your data allows.

When designing tables, remember that the default parameters used in standard software are not always the best choice for conveying technical information. Tables 5.5 and 5.6 contain the same data; however, Table 5.6 is much easier to read.

Table 5.5 Example of Table that Uses Default Parameters

Column heads are not spaced sensibly and are not in boldface.

Sieve Size	Weight Retained (g)	Percent Retained	Cumulative Percent Retained	Percent Passing
#4	54	10.2	10.2	89.9
#8	78	14.7	24.9	75.1
#16	125	23.5	48.4	51.6
#30	116	21.8	70.2	29.8
#50	88	16.6	86.8	13.2
#100	60	11.3	98.1	1.9
#200	5	0.9	99.0	1.0
Pan	5	0.9	100.0	0.0
Total	531	100.0		

Too many horizontal lines. Our eyes become distracted from the data.

Numbers and text are not centered.

All of these changes create a graphically simpler table that lets us focus on the data it presents rather than distracting us with meaningless lines and too much white space. Notice that the number and title of tables are placed on top.

Table 5.6 Example of Table with Modified Parameters

Sieve Size	Weight Retained (g)	Percent Retained	Cumulative Percent Retained	Percent Passing
#4	54	10.2	10.2	89.9
#8	78	14.7	24.9	75.1
#16	125	23.5	48.4	51.6
#30	116	21.8	70.2	29.8
#50	88	16.6	86.8	13.2
#100	60	11.3	98.1	1.9
#200	5	0.9	99.0	1.0
Pan	5	0.9	100.0	0.0
Total	531	100.0		

PRACTICE QUESTION: LABELING GRAPHICS

What is wrong with this title for a graph?

"Time vs. Temperature"

Suggested Answer

The title gives no more information than can be found by reading the axes of the graph. That information is already on the figure, so the title should add more information, such as the material involved (e.g., the temperature of what?).

PONDER THIS

5.4.4 A Word about Titles

Titles for graphics are often slapped on at the last minute without much thought. As such, in engineering documents, they are far too often superficial and do not help the reader understand the significant relationships or phenomena displayed by the graphic. A title that simply repeats the labels of the x-and y-axes adds nothing to the knowledge of the reader. Figure 5.21 depicts the decreasing velocity of a ball tossed into the air and has the title, "Velocity vs. Time." With this inadequate title as

KEY IDEA: Position each graphic as near as possible to the text it supports, but always after the first reference to it.

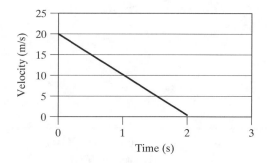

Figure 5.21
Velocity vs. Time [bad title].

a guide, we cannot understand why velocity is decreasing, because we have no idea as to what object's velocity is being measured. Here is a better, more descriptive title:

Figure 5.21. Velocity of ball tossed into the air (at constant acceleration of −9.8 meters/sec^2).

Another example of an inadequate title is shown in Figure 5.22, which depicts a graph showing air density as a prediction of the ideal-gas law. This *x*-vs.-*y* title simply repeats the labels on the figure without *interpreting* the data at all. A much better title would be the following:

Figure 5.22. Predicted relationship of density and temperature of air at standard atmospheric pressure

If any of your readers are likely to be nontechnical, you might want to add a caption after the title, such as the following:

Figure 5.22. Predicted inverse relationship of density and temperature of air at standard atmospheric pressure. Note that density tends to decrease with increasing temperature.

Captions are sometimes used in addition to titles. Captions can help the reader understand the data being presented; they often refer either to the way the data have been shown or to the conditions under which the data were collected. Let us look again at the title and caption from Table 5.4, which shows settlement data for piles. The caption tells us how to read the table for the greatest understanding. But, of course, this particular caption works only if the table is presented in color. Remember that *you cannot rely on your work always being reproduced in color in hard copy.*

Table 5.4. Comparison of settling for various types of piles.
Piles that exceed 0.25 inches of settlement are shown in red. Piles that are shaded in gray were discarded.

The punctuation of titles and captions has not been standardized across all engineering and scientific fields. Even the venerable *Chicago Manual of Style* does

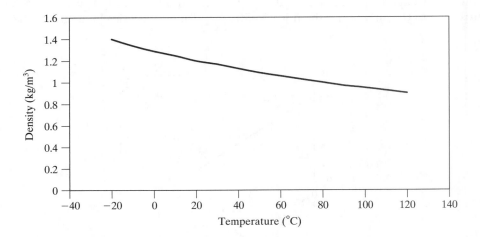

Figure 5.22
Density vs. Temperature
[bad title].

not indicate a preference for capitalization of titles and captions. Thus, you may either capitalize the first letter in all the words in your figure and table titles (not counting articles such as "a" and "the") or capitalize only the first letter of the initial word. If the figure has a title *and* a caption, the caption is most often treated as a sentence. You may have noticed that, in this chapter, all the figure titles have only the first letter of the initial word capitalized, whereas table titles have all "major" words capitalized [i.e., words other than conjunctions, prepositions, etc.]. In addition, the table titles in this chapter have no period, even if a caption follows, but figure titles have a period at the end, even if no caption follows. These decisions were made by the publisher, not by me. Most journals use the same format for both figure and table titles—the position (below vs. above the graphic), not the title's format, is what distinguishes a figure from a table. I suggest using periods only after the figure number (to separate it from the title) and not at the end of the title, unless it is followed by a caption. If there is a caption, then use a period to separate the title from the caption for both figures and tables. This latter treatment is recommended by the *Chicago Manual*. The best way for you to find out what to do is to read the journals in your field and notice how they handle figure titles and captions. Also, ask your instructor for guidance.

Guidelines for Graphics: *Labels*

- Label each figure or table clearly with a number and a title. For tables, the number and title are centered or left justified above the table. For figures, the number and title go beneath the figure. Your graphics program might have different defaults, but override them, unless your instructor says otherwise.
- Give pertinent details about the experiment either on the figure itself or in parentheses after the title.
- Create a title that draws attention to significant aspects of the illustration. Be more creative than using a simple "y vs. x" title.

5.5 INTEGRATING TEXT WITH FIGURES AND TABLES

The most important point to remember when integrating your text with your figures and tables is this: Each figure or table should be explained and interpreted in your text, and yet each should be able to stand by itself, to make sense even out of context. Your explanation of a figure or table should answer two questions sequentially:

1. What information does the figure or table contain, and (in many cases) how were those results obtained?
2. What does this information mean?

First orient the reader to the particulars shown in the figure or table (experimental details, values obtained, etc.); then fully interpret and discuss all changes, relationships, processes, or phenomena indicated by these data. If your figure or table is partially a repetition of information shown in earlier figures or tables, simply remind the reader of that information; do not repeat all of it. If the previous writing up to the point of any particular figure or table is good, very little new information should be necessary in order to answer the first question. You should focus instead on the meaning or interpretation of the information in the figure or table.

5.5.1 Using Graphics Designed by Others

KEY IDEA: Make your figures and tables right the first time, including the labeling and titling. It is easier to write and rewrite the text when you have final figures.

Sometimes, the graphic you want to use already exists, because the data it depicts did not originate with you. Sketches and photographs especially fall into this category, and they are readily available on the Internet. May you use a drawing created and published by someone else? For educational purposes (a report for class, for instance), the answer is a qualified "yes," as long as you acknowledge the source of the graphic. Figure 5.23 is an example of a picture used in a report comparing rocket engines. This drawing already existed on the NASA website and became an important tool in this student's research. There was no point in redesigning this visual depiction of the process by which a rocket engine produces thrust. So, the student provided the source of the drawing directly underneath the title.

> **When discussing technical graphics, here are some guidelines:**
>
> - Do not repeat in the text *all* of the data represented in the graphic (especially a table).
> - Do not continually say, "The figure shows.... " The emphasis should be on the information, not on the word "figure."
> - Do not fail to interpret the graphic and explain its significance.
> - Do not use "data" as a singular word. "Data" is a plural word; "datum" is the singular version.

It is a good idea to provide at least author information in the source citation. In this case, the author is the research center rather than an individual. The full citation should then go in the list of references at the end of the document. In this case, the full citation would include the URL (http://www.lerc.nasa.gov/www/pao/html/ipsworks.htm) and the date the student accessed the site, since the date of publication is not available on the Web page from which this picture came.

The site from which the Xenon Ion engine picture came is maintained by a public agency, so the images are usually considered to be public and therefore usable by the public, especially for educational purposes. The safest plan in borrowing an image, however, is to write to the authors or sponsors of the site (in this case, NASA) for permission to use the picture.

If you have designed your own graphical depiction of some data, but the data themselves came from another source, you do have an obligation to indicate the source of those data. For example, if you create an x–y plot to show the correlation between density of air and temperature at successive measurements, and the values for the plot came from a textbook, you should cite that book right under the title of your figure, as in Figure 5.24.

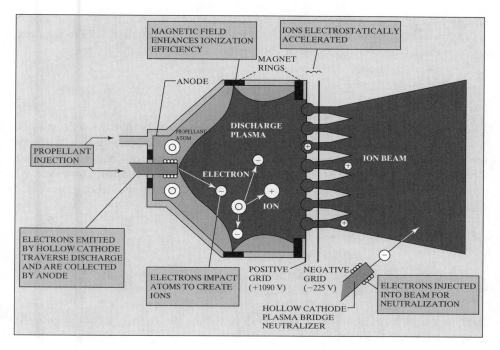

Figure 5.23
An internal schematic of the Xenon Ion engine. *Source:* NASA Glenn Research Center.

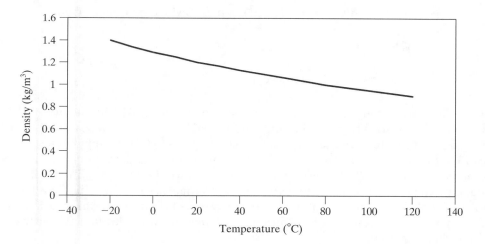

Figure 5.24
Relationship between density and temperature of air at standard atmospheric pressure. *Source of data:* Crowe, C.T., et al. 2001. *Engineering Fluid Mechanics*, 7th ed.

You do not have to cite all of the publication details under the title if those details are included in a reference list in your document.

5.6 MATHEMATICAL AND CHEMICAL NOTATION

Representing mathematical and chemical formulas and equations is easier today than ever before. MS Word has an Equation Editor that will render fractions and special relationships and offers a variety of special characters and Greek letters for equations. There are certain conventions for incorporating mathematical or chemical formulas into your writing. All equations are numbered sequentially in your text, and can thereafter be referred to in your text by the number:

$$Y = X^{2/3} + X \qquad (1)$$

$$E = mc^2 \qquad (2)$$

Einstein's simple equation is justly famous because it posits that two almost opposite entities, mass and energy, are related by a simple equation, and that the factor for converting one to the other is numerically enormous—the square of an already large number, the speed of light. For even simple equations, however, you must define all your variables and, if possible, give dimensions, either generally or specifically:

$$E = \text{energy (in joules)}$$
$$m = \text{mass (in kilograms)}$$
$$c = \text{speed of light in a vacuum (in meters/second;}$$
$$\text{approximately 300,000,000 m/s)}$$

In the case of Equation (1), you could define Y as the density of water in either general or specific terms:

$$Y = \text{density of water } (M/L^3) \quad [\text{mass divided by Liters cubed}]$$

OR

$$Y = \text{density of water } (g/c^3)$$

There is, of course, an International System of Units (SI) that defines many variables. This system is worldwide and is published by various organizations such as the National Institute for Standards and Technology (NIST): http://physics.nist.gov/cuu/Units/units.html. Consult the tables for notation of standard units, but, again, remember that not all of your readers or listeners will be familiar with all of the standard notations. It is safest to define all your variables somewhere in your document; for those variables that are used in several of your equations in a longer document (a technical report, for instance), define the variables in a Glossary or somewhere up front.

Be careful of using programming conventions to represent mathematical formulas in place of standard conventions; such use may lead to confusion. Consider, for instance, this formula generated from a spreadsheet application such as Excel:

$$Y = X^{\wedge}2/3 + X \qquad (3)$$

This notation could be misinterpreted to mean any of these three versions of the equation using conventional mathematical notation:

$$Y = \frac{X^2}{3 + X} \qquad (4)$$

OR

$$Y = \frac{X^2}{3} + X \tag{5}$$

OR

$$Y = X^{2/3} + X \tag{6}$$

Those familiar with programming languages will know that Equation (5) is the correct translation into conventional notation, but not all of your readers will interpret Equation (3) properly.

These same comments apply to chemical notation. Number the chemical reactions you represent. Define all symbols (even elements), unless you are writing for a technically advanced audience, and even then define symbols with any complexity;

As	Should be familiar to scientists and environmental engineers as "Arsenic"
CH_4	Might or might not be familiar to all engineers as a "molecule of methane"
CH_3COOH	Most likely needs defining as "acetic acid"

A courteous way to write the chemical formula for the breakdown of acetic acid would be:

$$CH_3COOH \text{ (acetic acid)} \longrightarrow CH_4 \text{ (methane)} + CO_2 \text{ (carbon dioxide)}$$

Greek letters should generally be defined, even those that are very common in chemical notation. If units are measured in μg/L, it is helpful to define the terms as micrograms/liter. And for some readers, it may be wise to define micro: $\mu = 10^{-6}$

Spreadsheet programs such as Excel generally follow the rule held by many engineering journals: use italic type for all letters representing variables except uppercase Greek letters, which should be in upright type.

SUMMARY

- Spreadsheet programs help you organize and synthesize numerical and other data that you collect. Use such programs to perform calculations on your data and, ultimately, to represent the data in chart form as a table or figure.
- If you experiment with the different chart forms available in Excel and other spreadsheet or graphing programs, you will learn which forms represent different kinds of data best. Bar charts and line graphs are good for showing changes in one variable over time; x–y and scatter plots are the best choices for showing changes in the relationship between two variables. Engineers use the latter types of graph most of the time.
- Readers need help in understanding the significance of the information in your graphics. Interpret and explain your graphics in words.
- Give proper credit when you use a graphic created by someone else. If possible, ask permission first.
- Use an equation editor and SI units to express mathematical and chemical formulae and equations. Define all variables.

PROBLEMS

5.1. Record the temperature in your bedroom over a 48-hour period. Then, make an *x–y* graph showing relationship between time of day/night and temperature. Answer these questions, and justify your choices:

What decisions do you have to make about displaying the data and labeling the axes?

Which kind of *x–y* graph did you choose (with line, with trendline, etc.)?

5.2. Take the same data on temperature of your bedroom as in Problem 5.1 and make another kind of graph or chart (other than an *x–y* graph). Try at least two types, such as these: a line graph, a pie chart, a bar chart, an area chart. Experiment to see which type of visual display looks best and displays the information best. Which type(s) do you think do *not* work well?

5.3. Test the graphs you produced in Problem 5.1 and/or Problem 5.2 on at least one friend. Which type of visual display works best and why? Ask your friends to explain their preferences.

5.4. Read an engineering article on a topic you are studying for one of your classes. Study the graphics and write a brief description of their effectiveness by answering the following questions:

Are the graphics easy to decipher and understand? Why or why not?

Do the graphics convey the information discussed in the text of the article, or do they introduce new data/information not completely discussed in the text?

Does each graphic contain enough details such that it could stand alone, without the accompanying discussion?

5.5. Look again at the graphics in the article you read in Problem 5.4. Which single graphic (figure or table) seems the best to you, and why? Discuss its virtues—what makes it clear and full of useful information? Can you point to design elements that help make this figure or table clear and useful?

5.6. Looking at the graphic of the Xenon Ion engine in Figure 5.23, write a paragraph or two of description of how the engine works. Does the graphic give you enough information to do this, or do you have to fill in some gaps?

5.7. True or false? Area charts (stacked charts) are the easiest to read of all types of charts.

5.8. True or false? Using color is always an effective way to highlight numbers on a table.

5.9. True or false? An acceptable strategy for presenting a set of data in reports is to include a selection of the data in a graphic in the body of the report and the full set of data in a table in the appendix.

5.10. Write at least one paragraph defending *one* of these positions:

(a) The y axis should always start at zero.

(b) Under certain circumstances, the *y* axis does not have to start at zero.

FURTHER READING

Cleveland, W.S. 1994. *The Elements of Graphing Data*. Hobart Press: Summit, NJ.

This comprehensive work covers every aspect of understanding how best to graph data. The discussion is high level, but easy to follow. There is more information here than most undergraduates will need, but everyone from beginning students to working engineers will benefit from the substantive discussions of how to make your data stand out and how to avoid ambiguity. Use the index to target answers to any questions you have about specific graphical elements. Error bars, for instance, can actually be ambiguous if the graph does not make clear whether they show the sample standard deviation (one use of error bars) or an estimate of the standard deviation of the standard mean (another use of error bars; see pp. 60–61).

Tufte, E.R. 1983. *The Visual Display of Quantitative Information*. Graphics Press: Cheshire, CT.

This beautifully produced book revolutionized the study of information display and raised the level of awareness of how important graphics are in understanding scientific and technical concepts. When Tufte could not get this book printed properly, he started his own publishing firm in Connecticut. Every page of this book has nuggets of gold, but you should look especially at Part I and the "Chartjunk" chapter in Part II.

Tufte, E.R. 1990. *Envisioning Information*. Graphics Press: Cheshire, CT.

This second major work by Tufte expands the variety of graphics examined, looking at every type of information display, including words themselves. There is a wonderful reproduction of a three-dimensional model published by Euclid in 1570.

Valiela, I. 2001. *Doing Science: Design, Analysis, and Communication of Scientific Research*. Oxford University Press: Oxford, U.K.

The first four chapters of this much-needed book cover how to obtain and analyze scientific information, as well as how to design one's research. Chapters 5–11 offer good guidance and examples for communicating those data effectively and accurately so that everyone, even the general public, can think critically about scientific information and make well-founded decisions.

Revising: When Will I Ever Be Finished?

Objectives

By reading this chapter, you will learn the following:

- the importance of reserving time for revision;
- how to develop peer-review groups;
- the benefits and shortcomings of grammar checkers;
- how to revise in manageable stages and correct for organization and content;
- how to edit paragraph by paragraph and correct for logic and flow;
- how to edit sentence by sentence and correct for grammar and style;
- how to use a checklist for finalizing and proofreading your written document;
- to consider your responsibility as an engineer when writing documents.

6.1 INTRODUCTION

The *hardest* part of writing is creating the first draft. When you initiate a writing project—whether it is a business letter or memo, a research proposal, a lab report, or a feasibility analysis—you are creating something out of nothing but the information you have gathered and the synthesizing powers of your own brain. This act of creation is exhausting, demanding, and finally, rewarding when you have a draft in hand. But the *most important* part of writing is *rewriting*. Rewriting is the multiphase sequence through which you strive to get your writing right, to make it meaningful to the reader and to make it accurately reflect your work and your thoughts. Your reader will never know all that you went through to produce any final document. And that is fine, because a good piece of writing is transparent. Its meaning just shines through the words, and the reader is not even aware of the craftsmanship. But to produce that transparent, seemingly effortless piece of writing, you usually have to craft many drafts. The trick is to make each revision count, so that you can see and track your progress in clarifying and completing the meanings you intend. And to track that progress, you often need the help of others.

Most people use the terms *rewriting, revising, editing*, and even *proofreading* to mean the same thing. Let us use *rewriting* here to mean the entire process of moving from first draft to finished product. *Revising* then means changing your document mainly for content and organization. *Editing* is a two-step process: correcting for logic and flow and improving your language sentence by sentence (this latter step is sometimes called "copyediting"). And *proofreading* is the final step in preparing a document, in which you correct mechanical errors.

6.2 WRITING IS REWRITING

Take a look at the following passage and determine whether it needs rewriting of some kind:

> Lilliputian Engineering gathered information about rope strength from several sources. First, Dr. Swift, a professor at Baker University, was consulted about the differences between ropes woven out of seaweed and grass. Then, we performed numerous lab experiments in which we determined the tensile strength of the individual fibers of the S-37, S-40, and J2000 ropes. **Resistance to discoloration was determined through phone interviews with company representatives**. After this, Lilliputian Engineering quantified internal rope stresses by hanging loads onto ropes of various diameters and measuring rope stretching.

This passage is not an example of terrible writing, but it is definitely writing that needs revision and editing. What are some of the problems? Well, for one thing, the sequence of investigation seems slightly askew. Surely, the researchers performed experiments on tensile strength *and* on internal stress before moving on to the interviews about discoloration. This report on rope strength seems to have been written in the order that the author remembered doing things rather than the actual, logical sequence of investigation or the order of importance of tasks. In addition, the sentence about phone interviews (in bold type) cries out for a citation (i.e., which company representatives were consulted?). And what about the imprecise nature of the words "numerous" (how many experiments is that?) in the third sentence and "this" in the last sentence? Also, do you not find it disconcerting that the sentences alternate between passive and active voice? In the third sentence, the writer states that "we *performed*" experiments, but in the fourth sentence, he or she says that discoloration "*was determined*." There is no rule about sticking with one verb form or the other throughout an entire paragraph, but the flip-flopping here seems unnecessary and awkward. In addition, if you really look closely, you will be bothered by the comparison in the second sentence: Are all the ropes woven out of both seaweed and grass, or are some ropes woven out of just one material?

So how do we fix all these errors? Well actually, most workplace documents undergo several rounds of editing and revising before they are considered final documents. Most technical editors advocate what is called "levels of edit" to break up the rewriting process into manageable chunks of effort and to ensure that any document works on all levels: content, organization, logic, and sense. This procedure may mean that you end up writing three or four drafts before achieving the final version. Sounds like a lot of effort, but remember that each draft is faster to produce, with fewer changes, and the result—a crisp, convincing, and thoughtful document—is well worth the effort. That document will get you what you want, whether it is funding for a new project, recognition for research well done, or a satisfied client.

Here is a checklist that presents important writing guidelines to help you revise and edit several times before handing in, circulating, or publishing any document.

KEY IDEA: Most workplace documents go through several rounds of editing and revising before they are considered final.

6.2.1 Rewriting: Guidelines for Revising, Editing, and Proofreading Your Writing

Whenever you finish a draft, put it aside for *at least a few hours*. Then follow this sequence:

1. Check the structure and content of your document:
   - Is the structure what the audience expects in a document of that type (report, proposal, etc.)?
   - Have you included all necessary information, and are your presentations of data as complete as you can make them? Have you indicated and justified omissions where you simply cannot be complete?
   - Do any passages include facts or figures that did not come from your original research? If so, would you consider that information to be general knowledge (in which case you do not have to cite references), or must you insert citations to outside sources of information?

2. Check the logic and flow of your ideas and information:
   - Is the logic based on the reader's frame of reference or yours?
   - Are there any passages that do not seem to follow logically from the preceding passage?
   - Does each paragraph begin with a topic sentence whose content is supported and expanded (through examples, description, etc.) throughout the rest of the paragraph?

3. Edit line by line for correct grammar and style: Check your sentences for sense, clarity, and precision.

4. Proofread for punctuation, spelling, words left out, spacing, and consistency of headings. Try reading the document aloud to "hear" any remaining errors.

The foregoing steps are generally sequential—step 1 usually comes before step 4—but they are also iterative. Proofreading, for instance, may reveal a missing piece of data that you thought you had included. When you make a substantive change (such as adding another piece of information), you will need to return at least to step 2 to check logic and flow, and then you will have to run through the remaining steps once again.

6.3 ORGANIZING A PEER REVIEW

KEY IDEA: When we read our own writing, we usually fail to see what we have left out.

Now, here is the million dollar question: Are you able to catch all the errors when revising something you wrote? If you are really honest, you will most likely answer, "No, probably not." It is extremely difficult for *any* writer to see all of his or her own errors. Because those sentences have come out of your head, you often continue to see what is in your head rather than what is on the page. Recent studies on human perception have shown that we often see what we expect to see even when it is not there or when something very unusual has taken its place. In an article in *The New Yorker* (June 11, 2001), writer Malcolm Gladwell gives an example from a psychological study (by Daniel Simons and Christopher Chabris) in which 50% of the observers of a videotape showing basketball players making passes failed to see the woman dressed up like a gorilla walk into the middle of the action. Those observers were so busy counting passes that they simply did not see a person in a gorilla suit standing in the middle of the basketball court. Gladwell's point in this article, about highway safety, is that research is beginning to show an astonishing insight: " . . . it is not just that our memory of what we see is selective; it is that seeing itself is selective" (p. 54). So it is no wonder that we do not see our misspellings or mistakes in grammar, but do see in our mind's eye that word we actually left out.

There are some proofreading strategies you can use to fool your mind into catching your own errors, but those strategies do not usually catch the more substantive mistakes that should be corrected. The best way to catch your errors in structure, logic, completeness, citation, and flow is to ask someone else to read and comment on your draft. Whom should you ask? Not your boss or your teacher (unless he or she asks to read your draft), but a peer—a fellow student or engineer. Preferably, your reviewer should be someone who is not familiar with the work you are writing about; that way, your reviewer will not substitute what he or she already knows about the subject for the information that is actually there on the page.

Now, it is not easy to ask someone to read a piece of writing that you know needs work. But think of the alternative: sending an unedited, incomplete, or just poorly written report or proposal to a decision maker (e.g., your boss, your teacher, or your client). It is better to risk embarrassment with a peer than run the risk of not getting a job or a decent grade. Besides, you can always offer to read your peer's drafts in return.

You can form a peer-review group with just two people or with many, at school or at work. Make a pact with each other that you will take the time to read carefully and comment on each other's drafts. You can do this reviewing easily online by using the Track Changes tool in MS Word. This feature shows all changes (additions and deletions) in a different color for each reviewer. You can then accept the changes you want and reject those you don't. The feature also lets your reviewer make comments and ask questions in the margins of your document. And you can always view the document with or without the changes shown. Here are some guidelines to follow when you ask for a review of your work:

PONDER THIS

How Do You Get the Most out of a Peer Review of Your Draft?

- Ask whether your reviewer wants to write down suggestions directly on your draft, to talk his or her comments through with you, or both. Ask whether he or she is more comfortable reading and commenting on the electronic or the hard copy.
- Come to an agreement with the reviewer on how much time he or she needs to review your draft; gently remind him or her if the deadline passes.
- Call your reviewer's attention to any critical elements on which you want him or her to focus.
- Do not expect a complete correction of all grammatical and sentence-structure errors. Your reviewer is probably not a grammarian, and what you really need are comments on the substance, completeness, and flow of your information.
- Receive all comments, whether written or verbal, in a positive, accepting spirit. Do not be defensive; think about how lucky you are to get this free critique that is bound to make your document better, if not perfect.
- Thank your reviewer, and always offer to reciprocate.

Here are some guidelines to follow when you review someone else's draft:

PONDER THIS

How Do You Give the Most when Reviewing the Writing of Others?

- Ask for the document in the form you are most comfortable working with: electronic or hard copy. If you know how to use the editing functions in your word processor (such as the Track Changes function in Microsoft Word), use them to make suggestions for revision. Consider learning about these editing functions if you are not already familiar with them.
- Try to comment on what is good as well as what could be better.
- Do not use language that is personally critical: "I think there are a lot of sentence-level errors in this passage," *not* "Your writing really stinks here."
- Focus on the big picture—the clarity, structure, and flow of the document—rather than on catching every word error. You do not want this task to be so much work that you end up hating the draft *and* its author.
- Whenever possible, offer to meet with the writer and go over the document together. It is easier to explain some things in person than in written comments.
- Use a positive, upbeat tone in your verbal comments, even if you see lots of errors. Be patient in your explanations. Writers are very sensitive about their own creations and often have a hard time seeing another way of saying something.

The process of peer review has recently been made much easier by the invention of collaborative editing tools such as MS Word's Track Changes. Using this function, you can add or remove text to the original document and make comments in the margin, without erasing the original wording. The original author can decide whether to accept or reject any proposed change, and the comments/questions help him understand the confusion or suggestion of the peer reviewer. Using Track Changes is like having a multi-layered conversation among versions of a document. The changes and markings may seem overwhelming sometimes, but they always end in producing a better piece of writing. And you can always "turn off" the changes without accepting all of them, by choosing "Final" as the view you want. Most co-authors would be lost these days without the benefit of a function such as Track Changes, which literally tracks the changes made and never lets any of them go until you say so. If you and your reviewer have the time, talk together about the suggestions and questions he or she has; if you don't have the time, read the comments carefully before deleting them.

Remember that in the workplace, peer review does not necessarily take the place of a final review by a professional editor or technical writer within the organization. The better the draft you give the technical editor, the more he or she can focus on polishing your writing instead of trying to make substantive changes in subject matter on which he or she is not an expert.

Some school or work groups use guided questions to stimulate and standardize the peer-review process. Rather than asking reviewers to completely edit your work, it is a good idea to ask them to focus on certain issues that tend to arise in your particular document. For example, say that you have written a research proposal for your company. In this company, proposals of this type usually outline methods for conducting a feasibility study of solutions to the client's need or problem. Budgets included in these types of proposals should make clear that the costs are for conducting an investigation of possible solutions, not for implementing a chosen solution. In the following set of questions for peer reviewers of your proposal, the words with initial capital letters are headings of required sections in the proposal:

Sample Peer-Review Questions for a Research Proposal

1. Does the Summary seem complete? If the Summary you are reviewing seems incomplete or unpersuasive, let the author know, and offer a suggestion for improvement.

2. What evidence do you find in the Problem Definition that the author truly understands the client's problem?

3. Does the Scope of Project state the specific objective(s) of this project? Does it outline the alternative solutions that will be investigated and the methods of study that will be used to investigate and evaluate them? Note on the document where these requirements are fulfilled.

4. Has the author indicated that certain evaluation criteria are more important than others? If the criteria are not prioritized, can you suggest how they might be?

5. Are the steps outlined in the Proposed Procedure specific enough? If so, explain how each step follows logically from the others? If not, please indicate where the author should provide more detail.

6. Use the Proposed Procedure section to check the Schedule. What are the tasks listed in the Schedule, and are they sufficiently explained in the Proposed Procedure? Do you have any suggestions for revision?

7. Comment on how realistic the Budget is. Does it reflect the cost of the author's conducting the study, *not* the cost of implementing a solution to the client's problem?

8. Do you think the proposal is persuasive? Do you think the proposal would convince the primary audience (the client) to fund this project? Why or why not?

9. Where can you see any evidence in this proposal that it is written also for the secondary audience, those other interested readers (such as advisor or owners or financial managers)?

As an individual you can of course, ask your reviewer to focus on certain issues that tend to arise in your own writing. For instance, if you know that you have trouble getting to the point, be sure to have someone check your statements to make sure you have clearly explained at the beginning of the document what the project is all about and what the document intends to do (propose something, conclude something, etc.). In other words, you want to ask your human reviewers to do what computer-based grammar checkers cannot do: check for clarity, logic, consistency with company standards, and completeness of information.

6.4 ABOUT GRAMMAR CHECKERS

The grammar-checker function included in most word-processing software is, on the whole, a boon to clear writing. Grammar checkers highlight for us the phrases and sentences that may contain errors in grammar or style. Figure 6.1, for instance, is the screen that pops up when I do a grammar check in Microsoft Word on a funky sentence I have written critiquing another author's writing.

I knew the sentence was "funky" because Word had underlined it in green. To find out more specifically what is wrong with it, I chose the Spelling and Grammar function from the Tools menu. The checker considers my question to be a fragment for a couple of reasons, probably: The question starts with "and," and I have buried a question within a question (in parentheses). The first problem is arguable: Some

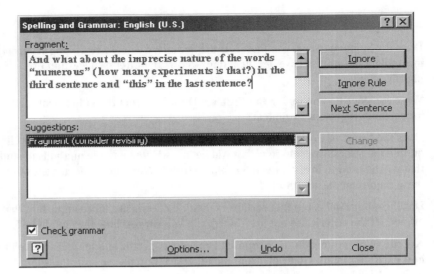

Figure 6.1
A sample screen from the grammar-checking tool in Microsoft Word.

grammarians believe it is never proper to start a sentence with "and" or "but," and some think that it is OK to do so. In fact, the 15th edition of the *Chicago Manual of Style* (2003) says it is perfectly fine to start a sentence with a coordinating conjunction ("and" and but" are the most common), but traditional composition guidelines die hard. Without going into more details, I suggest a middle-of-the-road course that limits your use of "and" or "but" as a sentence starter to those infrequent occasions when the link (similarity or dissimilarity) to the previous sentence is more important than any other meaning in the current sentence.

The second problem—burying a sentence within a sentence—is a more serious issue and should never be done in formal or even semiformal technical writing. I chose to leave that sentence as it is because I want to be informal in the style of this book. Also, when I am picking apart a passage to explain stylistic nuances, I have to work harder to get your attention. A question buried within a question may have gotten your attention. In any case, I considered what the Word grammar checker indicated and then used my own judgment to decide what to do.

KEY IDEA: You cannot rely on grammar checkers alone to catch spelling and grammatical errors because the software does not read for *meaning*.

As long as you use your own judgment in conjunction with your word processor's promptings, grammar checkers are helpful in catching those grammatical and stylistic errors we tend to make all the time (such as overuse of passive voice and disagreement of subject and verb). In addition, the spell checker is good at catching those words we tend to misspell. I can never remember, for instance, whether "attendence" or "attendance" is correct. But Word's spell checker tells me immediately that the first spelling is incorrect by underlining it in red. However, there are many misspelled words that Word cannot catch because *Word does not read for meaning*. Word processors thus far do not have artificial intelligence, so they cannot read for sense; for example, they cannot know whether the correct word for you to use in a particular sentence is "from" or "form." Both spellings are words in the dictionary, and thus only a being with a certain level of cognitive reasoning skills can tell you whether you want to write: "The data were extrapolated form many sources" or "The data were extrapolated from many sources." That being is usually you, unless you have a careful peer reviewer or the luxury of a technical editor. My word processor did not underline "form" in any color in the previous sentence, even though it was incorrect in the context.

Practice Question: Grammar Checkers

Type this sentence into your word processor, and then run the grammar checker on it:

The director explained the environmental process review guidelines.

Did the checker make any suggestions? If not, can you pinpoint at least one problem with the readability of this sentence? Is anything unclear?

Suggested Answer

The noun phrase, "environmental process review guidelines," is rather difficult to understand precisely. Is the director explaining a process for reviewing environmental guidelines? Or is she explaining the guidelines for reviewing environmental processes? If your grammar checker did not ask for clarification here, then you will have to catch these confusing noun phrases for yourself.

6.5 REVISING FIRST FOR STRUCTURE AND CONTENT

Revising for structure and content is often difficult to separate from editing for logic and flow (the next phase). After all, if you discover that some content is missing (some specific test results, for instance), then your conclusions will not flow logically from the data presented; there is a problem with content *and* with flow, and of course you will have to restructure that section. Do not worry about trying to separate out the two phases; just proceed from the macro to the micro as you revise and edit. Start by looking at the entire document (remember to check your table of contents for consistency with headings in the body of the document), move fairly quickly from section to section, and keep in mind the big picture. Ask yourself these questions as you read:

1. Does this draft conform to standards and expectations for this type of document in my company, agency, or classroom?
2. Have I highlighted the really important points, even if that means I repeat them in different places in the document?
3. Do I introduce the document and each section properly so that the reader knows what to expect next?
4. What do I want the reader to get out of this particular section?

To answer the first question, make sure that your draft includes the sections (with appropriate headings) expected by your client, boss, or teacher. If you do not know what is expected, go and find examples of similar documents. Certainly, you may have to add or subtract certain sections as you realize that they do or do not apply to the particular project you are writing about, but make these changes carefully and justify them to yourself. For documents that are longer than four pages, adding a summary at the beginning is always a good idea. For longer reports, this section is called an executive summary. For most documents, a conclusion, whether labeled as such or not, should summarize and highlight important points for the reader.

As an example of how a good piece of writing answers those four questions, let us examine this summary for a proposal:

Proposal to Evaluate Wall Designs for Apartment Sound Control

Summary

ERC Properties, an Austin apartment management company, is searching for a wall design to control sound in their new apartment complexes. In their existing complexes, residents have complained about the insufficient sound control provided by the common wall between two apartments. These high levels of noise have convinced some residents to move to other apartment complexes, forcing ERC Properties to search for replacement tenants. To avoid losing residents in their future complexes, ERC has determined to improve the sound control. Therefore, they have asked Firm Ideas to evaluate different wall types and materials that would provide better sound control in the new complexes.

Firm Ideas has begun the process of choosing and evaluating *wall designs* for *controlling noise* between apartments. This proposal outlines the *initial designs* and the methods for evaluating them. These six initial designs include two types each of single-, staggered-, and double-stud walls. Each *wall design* will be evaluated based on the wall's performance, material and labor cost, construction time, and usage limitations. After the evaluations are complete, Firm Ideas will recommend the *most suitable wall design* to ERC Properties.

The wall-design evaluations are expected to take 10 weeks. A progress report of this project will be available on March 31, 2003, and a final report will be submitted to ERC Properties on May 1, 2003. The estimated cost of Firm Idea's services on this project is $1,750.00.

> checklist item 2:
> The most important points in this proposal have to do with the *problem* of controlling sound between apartments and with the various *solutions* that involve redesigning the walls. Repeating the words "sound control" emphasizes those points.

> checklist item 4: Methodology statements convince the *advisor* reader.

> checklist item 4: Cost and time estimates are important for the *decision-making* reader.

Most proposals start with a summary, so the reader's expectations are being fulfilled here (item 1 of the foregoing checklist). Notice that this summary also answers the most important questions for the client and thereby fulfills items 2 (highlighting important points) and 4 (educating readers). Of course, different readers of this proposal will have somewhat different concerns: The *decision-making* reader will want primarily to know how quickly and at what cost the sound-control problem can be solved. The *advisor* reader will want to know how carefully the writers evaluated various solutions. At the beginning of a report or proposal, however, all readers ask themselves the following questions:

What is the purpose of the document I am reading?
What is the purpose of the proposed project?
How long will this project take, and how much will it cost?

The proposal summary answers those questions in three short paragraphs. In fact, the summary has fulfilled item 3 in the foregoing checklist (letting the reader know what to expect in the section) by letting the reader know what to expect in the entire document. For other sections in this proposal, how do you introduce them so as to preview their content for the reader? Here is the beginning of a section called "Solution Criteria":

> Each of the six wall designs will be evaluated according to four criteria: performance, cost, time to construct, and usage limitations. These criteria are explained next in prioritized order.

Those two sentences tell us exactly what to expect next and, indeed, from the whole section.

Checklist for Revision of Structure and Content

1. Ensure that the text conforms to structural standards for the particular type of document.
2. Check that the document highlights important points (perhaps with graphics).
3. Make sure that the document introduces each section by making the section's purpose clear.
4. Ensure that each section fulfills a purpose for the various readers of the document.

6.6 EDITING FOR LOGIC AND FLOW (MOSTLY PARAGRAPH LEVEL)

Making sure that you let the reader know what to expect next is the task that connects reading for content with reading for logic and flow. Since the most important location in any document is usually the beginning, be sure to check the first section (whether a summary, an introduction, or just the first paragraph of a letter) most carefully for logic and flow. Ask yourself if the text clearly conveys the purpose of the document and what you want the reader to get out of the section. Make sure the opening statements follow a logical order and present the most important information first. If we look again at the first paragraph of the proposal summary, we see that it follows a sequence that moves from company *need* to underlying company *problem* and *source* of that problem:

ERC Properties, an Austin apartment management company, is *searching for a wall design to control sound* in their new apartment complexes. In their existing complexes, residents have complained about *the insufficient sound control* provided by the common wall between two apartments. These *high levels of noise* have convinced *some residents to move* to other apartment complexes, thus *forcing ERC Properties to search for replacement tenants.* To avoid losing residents in their future complexes, ERC has determined to improve the sound control. Therefore, they have asked Firm Ideas to evaluate different wall types and materials that would provide better sound control in the new complexes.

The client's *need* (what this proposal is responding to).

Source of noise problem.

Source of three problems for client:
1. Tenants are moving out.
2. ERC must search for new tenants.
3. Future tenants may move, too.

After checking the beginning section carefully, read your document paragraph by paragraph. Reading at the paragraph level often becomes intermixed with reading at the document level, but try to focus on flow and transitions, and do not worry so much about the big questions of purpose and content. Let us look at an *earlier* version of the paragraph we just examined. Do you see places where the logic is not as tight as it could be, where a step seems to be missing in the explanation presented?

First Draft of First Paragraph of a Research Proposal

ERC Properties, a central Texas apartment management company, is searching for a wall design to control sound transmitted through shared walls in an apartment complex. ERC Properties has received complaints from residents of transmitted noise through adjacent units in several of their apartment complexes. *These high levels of noise have convinced some residents to move to other apartment complexes, thus forcing ERC Properties to search for better sound control.* Therefore, they have asked Firm Ideas to evaluate different wall types and materials that would provide better sound control in the new complexes.

Take a close look at the third sentence. Is the cause-and-effect relationship presented there entirely clear? Actually, there are two cause-and-effect relationships—an implied one in the first half of the sentence (residents are moving because of the "high levels of noise") and a direct one in the second half (residents are moving; "thus," ERC is searching "for better sound control"):

> *Implied* cause and effect

These high levels of noise have convinced some residents to move to other apartment complexes, thus forcing ERC Properties to search for better sound control.

> *Stated* cause and effect: residents' moving means ERC must search for better sound control

KEY IDEA: Read your own work for logic and completeness. Are you breaking causal relationships down into logical steps that don't make unwarranted assumptions?

But does it necessarily follow that residents' leaving means that ERC must search for better sound control? There is actually a step missing: Residents are leaving; thus, ERC is forced to find new tenants, and *in order to find and keep new tenants*, ERC must improve sound control. You help the reader understand more quickly if you break causal relationships down into the simplest chain possible, which may mean adding more links to the chain. When you add more links to a cause-and-effect chain, you usually need to keep sentences as short as possible; thus, you often end up crafting more sentences to clarify the cause-and-effect relationships. Check out this revision, in which the original sentence is now broken into two sentences:

> These high levels of noise have convinced some residents to move to other apartment complexes, thus forcing ERC Properties to search for replacement tenants. To avoid losing residents in their future complexes, ERC has determined to improve the sound control.

In the first sentence, the cause-and-effect chain has two large links (or steps): The high levels of noise cause residents to move, and residents' moving causes ERC to search for replacement tenants. Notice the clarification here: ERC must "search" for "replacement tenants," and that search has led them to search for ways to improve sound control in future buildings. In the second sentence, the chain moves backward, presenting the desired effect first—to avoid losing tenants in the future—and the cause second—ERC's projected improvement of sound control. This way of explaining things has the added benefit of presenting ERC's search for sound control as the cause of the current proposal writers' investigation as well. The next sentence practically writes itself (as yet another effect of the cause presented in the previous sentence):

> Therefore, they have asked Firm Ideas to evaluate different wall types and materials that would provide better sound control in the new complexes.

What else needs clarifying in this first draft of the proposal's first paragraph? If ERC wants new tenants to rent their apartments, and they have determined that inadequate sound control is the problem that keeps tenants away, then we would expect that ERC would try to renovate their existing apartments to provide better sound control. But our expectations are wrong; we are misled for most of the paragraph. We learn only in the last sentence, second-to-last word, that ERC is not in fact going to renovate existing apartments: They are planning on incorporating better sound-control materials in their *new* apartment complexes. The edited version clarifies this difference between existing and new apartments simply by using the words "existing," "new," and "future" earlier in the paragraph.

First Paragraph of Research Proposal Edited for Logic and Flow

ERC Properties, an Austin apartment management company, is searching for a wall design to control sound in their *new* apartment complexes. In their *existing* complexes, residents have complained about the insufficient sound control provided by the common wall between two apartments. These high levels of noise have convinced some residents to move to other apartment complexes, thus forcing ERC Properties to search for replacement tenants. To avoid losing residents in their *future* complexes, ERC has determined to improve the sound control. Therefore, they have asked Firm Ideas to evaluate different wall types and materials that would provide better sound control in the *new* complexes.

Checklist for the *Logic-and-Flow* Level of Editing (Paragraph Level)

1. Ensure that the document uses organizing patterns such as cause-and-effect or problem–solution structures.
2. Within that pattern, each sentence of the paragraph flows logically from the previous sentence.
3. Make sure that links in cause-and-effect chains are clearly delineated. There should be no more than two links per sentence.
4. Ensure that the topic sentence introduces the theme of the paragraph and that it makes a logical link with the final sentence.
5. Make sure that no sentence in a paragraph is more general than the paragraph's topic sentence.

6.7 EDITING FOR SENSE AND MEANING (MOSTLY SENTENCE AND WORD LEVELS)

After we edit a document for logic and flow, we should take one more look at it in order to concentrate on editing it at the sentence level, checking for sense and meaning and making sure that every word is as precise as possible and the grammar is correct. Consider once again the following paragraph, where each sentence has been numbered for the purposes of our discussion:

> [1] ERC Properties, an Austin apartment management company, is searching for a wall design to control sound in their new apartment complexes. [2] In their existing complexes, residents have complained about the insufficient sound control provided by the common wall between two apartments. [3] These high levels of noise have convinced some residents to move to other apartment complexes, thus forcing ERC Properties to search for replacement tenants. [4] To avoid losing residents in their future complexes, ERC has determined to improve the sound control. [5] Therefore, they have asked Firm Ideas to evaluate different wall types and materials that would provide better sound control in the new complexes.

KEY IDEA: Watch out for undefined pronouns, such as "their," "it," "this."

When you are editing sentence by sentence, you need to slow yourself down, but keep reading for sense. (In proofreading, the final phase, you focus more on mechanics than sense.) One way to slow down and really see what is written is to read the text out loud to yourself. This technique has worked for many writers. The trick is to keep reading aloud and not lapse into reading silently. The technique depends on your ear hearing what your eye no longer sees (e.g., a word left out, an overly wordy sentence, or an ambiguous pronoun).

A possible problem in the foregoing paragraph occurs in sentence #1 with the pronoun "their." "Their" refers to ERC Properties, a company, which is a *collective* noun, meaning it can be either singular or plural, depending on the meaning. Here, the company is presumably acting as a single entity, and indeed the singular verb "is" describes what the company is doing ("searching"). Why then, does the writer switch and use the plural pronoun "their" in the second part of the sentence? Well, the singular form of the pronoun that would apply is "it," since we do not usually refer to companies as "he" or "she." But "it" does not seem quite right either, since we know that the company is composed of human beings making decisions and doing the sort of cognitive work that an "it" cannot do. So, we often cheat a bit and use the plural pronoun—a form of "they"—to indicate a singular entity composed of more than one human. The practice is so common by now that it does not seem worth worrying about, and the fixes are also awkward; it is easy enough to say "ERC Properties … *are* searching," but it is a bit more difficult to use a plural verb with the shortened form of the company's name, as in, "ERC *have* determined to improve." In sentence #1, every reader will know whom you are indicating by "their," but be careful when there is more than one plural noun in the sentence. Look at sentence #2: "*their* existing complexes" could refer to "residents," since that is the plural noun closest to the pronoun "their." You would be wise not to use "their" there. So, let us change that sentence by removing "their" and repeating a shortened version of the client company's name:

In existing **ERC** complexes, residents have complained about the insufficient sound control provided by the common wall between two apartments.

You might also want to change the "their" to "the" in sentence #4.

Unnecessary wordiness is a common problem with almost everyone's writing. The paragraph here has become quite streamlined and focused through the process of editing for flow, so we do not see a lot of extra words. But how about the word "thus" in sentence #3? It is a common and useful transitional word, but we do not really need it here because we have the participle "forcing" to explain the consequence of residents' moving out. Rereading this paragraph multiple times enables us to see another wordiness issue: The word "complexes" is used in every single sentence! It is an important word to use, because "apartments" or "buildings" does not mean quite the same thing, and we want to use consistent terminology, especially when focusing on the client's problem. If we examine each sentence, however, we can see that "apartment complexes" is probably not necessary—or even accurate—in sentence #3. We can just say "convinced some residents to move out." In addition, we may not need the introductory phrase "In existing ERC complexes" in sentence #2; we could just make that information part of the main clause: "Residents of existing complexes have complained. ... " That way, we can remove a comma and get right to the main subject of the sentence: "residents."

So here is our new, streamlined version of the paragraph:

ERC Properties, an Austin apartment management company, is searching for a wall design to control sound in their new apartment complexes. [2] Residents of existing ERC complexes have complained about the insufficient sound control provided by the common wall between two apartments. [3] These high levels of noise have convinced some residents to move out, forcing ERC Properties to search for replacement tenants. [4] To avoid losing residents in planned future complexes, ERC has determined to improve the sound control. [5] Therefore, they have asked Firm Ideas to evaluate different wall types and materials that would provide better sound control in the new complexes.

Actually, the previous version of this paragraph that had been revised for logic and flow had very few sentence and word-level issues. This degree of polish is a common reward for paying attention to each level of edit as you work your way through the rewriting process.

At this sense-and-meaning level of edit, though, you want to remember what you know about your own stylistic and grammatical foibles. What have writing teachers commented on in the past? Do you tend to use a lot of passive voice? Are your sentences described as "wordy" on returned papers? If punctuation is a particular problem for you, you will need to pay attention to it at every level of editing, because every time you rephrase a sentence, the punctuation has to be rethought as well. You can use the Find function of your word processor to help target word and punctuation mistakes you tend to make. For example, because the apostrophe is usually used to indicate the possessive, many people use "it's" mistakenly. "It's" means "it is," not the possessive of "it," which is indicated by "its," as in "its function" (the function of "it"). I usually set my word processor to look for both "it's" and "its" so that I can check for correct usage. Similarly, if the semicolon is still an untamed beast to you, search for words such as "however" and "therefore." If one of those words occurs in the middle of a sentence, see whether each set of words on either side could form a complete sentence. If so, you probably need to put a semicolon before the "however" and a comma afterward.

You want at this point also to check that your numbers are not only accurate but also refined to a reliable level of precision. If a probe is precise only to one decimal point, don't report the measurement to finer precisions of two or more decimal places. On the axes of graphs and in tables, make sure the scales and the reported numbers are shown to the same number of decimal points. Think always about the scale at which you are working; don't, for instance, report financial-analysis results to the penny if the figures involved are in the thousands or millions.

Checklist for *Sense-and-Meaning* Level of Edit (Sentences and Words)

1. Ensure that sentences are linked together with transitional words where necessary.
2. Make sure that words are not ambiguous in meaning
3. Check that technical terms and keywords are used consistently and not changed just for variety's sake.
4. Consider suggestions by your computer's grammar checker.
5. Ensure that sentences have been checked for these common problems:
   - too much passive voice
   - ambiguous pronoun reference
   - subject–verb disagreement
   - wrong word: form/from, their/there/they're, cite/site, to/too/two, etc.
6. Check that figures and tables have meaningful titles and adequate labeling.

Use the computer's "Find" function to search for your common errors

---

Practice Question: Computers and Revision

If grammar checkers are still so unreliable, how do you use your word processor to pinpoint problems at the sentence level?

Suggested Answer

Use the Find function to search for particular words and phrases that you know you tend to misuse. If your instructor is a stickler for using "data" as a plural word, then search the document for "data" and make sure every verb connected to that noun is plural.

6.8 PROOFREADING FOR FINAL COPY

At this final phase in the process of improving your document, you are checking for mechanical errors, which include the following:

- spacing problems (between both lines and words)
- mistakes in numbering (sections, headings, pages, figures, and tables) mistakes in mathematical notation
- missing words and marks of punctuation (especially the right-hand quotation mark or parenthesis)
- misspellings
- incorrect calculations or reporting of numbers

Checking your numerical calculations and being consistent in your notation are two important proofreading tasks. If some of the numbers in your table are reported to the first decimal point, chances are they all should be. Do not include variations such as these in the same column or row:
1.22, 3.2, 4, 4.02

Report your numbers consistently:
1.22, 3.20, 4.00, 4.02

And as you check the numbers in your figures, tables, and text, be sure that the precision reported truly represents the actual reliability of the numbers. A probe, for example, may be precise to one decimal point; do not report the measurements to finer precisions of two or more decimal places. Similarly, do not report the results of financial analyses to the penny, if those results involve figures of thousands or millions of dollars.

Some writing experts suggest reading backward in order to proofread without getting caught up in the sense of the sentences. You could read the entire document backward, word for word, or you could read each sentence forward, starting with the final sentence and moving backward sentence by sentence through the document. I have to say that I find this method unbearably tedious and cannot make myself do it. I read forward, but very slowly, paying little attention to sense and lots of attention to the look of the words on the page and to sequences of any kind: numbering, pairs of punctuation marks, and all those little details that are difficult to pay much attention to but make the difference between reader understanding and confusion.

Whatever proofreading method you use, make sure that you allow time after editing for flow and sense before you attempt final proofing. Of course, if you catch a problem with sense or flow during proofing, you will want to correct it, but you are likely pretty tired at this point, and you may no longer be the best judge of what flows or makes sense. If you have already edited for sense and meaning, then clean up the mechanical errors of punctuation, numbering, spacing, and spelling, and you will have a professional-looking document, even if a few language errors remain. Chances are good, in any case, that those remaining errors are matters of style and therefore questionable and not necessarily rule bound.

Speaking of getting tired, remember the reality of the writing process: Writing is tiring. When you find a particular mechanical error (such as no right-hand quotation mark), you are likely to find others nearby. Simple mistakes tend to cluster in our documents at the point at which we got tired, and therefore sloppy.

KEY IDEA

Suggestions for Powerful Proofreading

- Allow time after editing.
- Check numbering.
- Check whatever comes in pairs (e.g., quotation marks and parentheses).
- Check numerical calculations.
- Remember that mistakes tend to occur in clusters.
- Turn the document upside down to check spacing.
- Consider reading backwards.
- Use the spell checker!!!

Practice Question: Proofreading

Proofread the following passage to see how many errors you can catch and correct:

Everyone had their own idea about how the software would be developed, what platform would run on best, and what hardware would make it most scalable. When the company folded, the former CEO says "what should have been a brilliant new product for pricing derivatives is only a half completed bundle of program code. The CEO is now seek funding for a new start-up venture.

Suggested Answer

Does your proofread version look like this?

Everyone had *his or her* own idea about how the software would be developed, what platform *it* would run on best, and what hardware would make it most scalable. When the company folded, the former CEO *said*, "What should have been a brilliant new product for pricing derivatives is only a *half-completed* bundle of program code." The CEO is now seeking funding for a new venture.

6.9 ETHICS AND WRITTEN DOCUMENTS

As a final test, be sure that the document you have just finished is truthful and does not leave out any information that would be important for the readers to know. You probably know by now that ethical dilemmas arise as part of everyday engineering practice and provoke questions such as these:

1. Because no one else is around and the work needs to proceed, should you sign off on a document certifying work for which you don't have expertise?
2. Should you cut corners in your design only because it's cheaper that way?
3. Should you do whatever the client wants, even if you think it's ethically dubious?

Some of these questions are easier to answer than others. The answers to questions 1 and 3 should probably be no; the answer to #2 may be more complicated, depending on what "cut corners" means. As long as you consider safety first, you do need to consider cost when designing engineering solutions.

What you may not yet be aware of is that many of the ethical dilemmas you will face involve communication:

- Should you omit data from your report because it doesn't fit with the trend of the rest of the data?
- Should you bury the recommendations in the middle of the report because your boss is concerned about liability issues if the recommendations don't work?

As an engineer, your primary responsibility is to the people who will use or be impacted by your work. And sometimes that responsibility comes into conflict with your responsibility to your employer. Whenever you have questions about the right course of action, consult the engineering codes of practice. The National Society of Professional Engineers has one, and so do most of the professional engineering associations (American Society of Civil Engineers, American Chemical Association, etc.). Bookmark their websites and become familiar with their principles. As a new engineer, it is also a great idea to find a mentor in your company or organization: someone who may or may not be your boss but whom you trust and who seems to take a positive interest in you. Discuss challenging situations with your mentor.

Be sure to give credit to the work of others in all that you write and design. The NSPE Code of Ethics states this requirement as a professional obligation:

III Professional Obligations

9. Engineers shall give credit for engineering work to those to whom credit is due, and will recognize the proprietary interests of others.
  a. Engineers shall, whenever possible, name the person or persons who may be individually responsible for designs, inventions, writings, or other accomplishments.

Do not plagiarize, even unintentionally. If you use someone else's words in your writing or state ideas that are not common knowledge and are not part of your original research, then you must credit the author of those words or ideas. Consult guidebooks such as the *Prentice-Hall Reference Guide to Grammar and Usage* for examples of how to cite sources properly. If you reuse your own words (from one assignment to another, for instance), be sure to check with your instructor about whether that reuse is acceptable. Ethics researchers such as Michael Loui, have stated that quoting yourself is not plagiarism, but different disciplines have different conventions and expectations about citing your own previous work. Always err on the side of over-citing rather than under-citing.

Ethical dilemmas arise as part of everyday engineering practice.

Many situations involve communication:
  —Being asked to distort results in reports
  —Being asked to de-emphasize recommendations
  —Being asked to withhold information from clients
  —Being asked to break oral or written agreements
  —Being asked to distort reported data or omit data.
  —Failing to give credit in published reports.

KEY IDEA

A final consideration that is sometimes forgotten by engineering authors is whether or not the finished document is true to the engineer's ethical canon requiring him to "hold paramount the safety of the public." Every so often, interests other than engineering and technical ones become predominant in the workplace, and accuracy and truth take a back seat to what is economically expedient or what is less "risky" for the company.

SUMMARY

- Revising your writing may be the most important part of the writing process. Approach revising in stages, focusing first on rewriting for structure and content.
- Edit also in stages, focusing first on examining paragraphs and then on correcting sentences for grammar and clarity.
- Do not rely on electronic grammar checkers to catch your sentence-level errors. Use the checkers with caution, and make sure you have a good style guide at hand.
- Asking others to review your writing before it is finalized is a great idea. Form a peer-review group with other students or coworkers. Receiving a peer review, however, is no substitute for doing a final edit yourself.
- Learn to leave time for proofreading, and adopt the strategies that work best for you. Be prepared to be professionally responsible for the truth, accuracy, and usability of your document.
- Check your document for truthfulness and make sure you have given proper credit to the words and ideas of others.

PROBLEMS

6.1. Take a draft of a current writing assignment you have been given and edit it yourself. Time yourself and see how long it takes you to produce a second draft. Write a brief explanation of how long the second draft took to produce, how difficult the edit was, and whether you think the assignment is now in final form. Could you feel good about handing in the second draft?

6.2. Take the same draft as in Problem 6.1 (the first draft, not the revision) and give it to a friend to edit. Ask your friend to read as carefully as possible, not to be afraid to criticize, and to write down all the errors he or she can see. Ask your friend to indicate where the writing gets confusing.

- Now, if you have already edited the draft yourself, compare the two edits—yours and your friend's. How are they different? Write a brief description of the differences.
- If you have not already edited the draft yourself, write a list of the mistakes and problems your friend encountered in editing your draft. What surprises you?

6.3. Ask the friend who edited your draft in Problem 6.2 to talk to you about his or her suggestions. Is it easier to understand someone's criticism when they explain it to you verbally? On a scale of 1 ("fabulously helpful") to 10 ("a total waste of time"), rate the usefulness of having someone else edit your work. Which was more useful: the written comments or the verbal comments?

6.4. Given the way most people read, the most important part of a document is which part?

The beginning

The middle

The end

6.5. Under what circumstances would you want to present background information about a problem before stating the solution? Under what circumstances would you want to present the solution before the details of the problem?

6.6. True or false? In the stages of rewriting, editing for sense and clarity should come before editing for structure and content.

6.7. True or false? Once I become a good enough writer, I will be able to write two-page (or even longer) documents in one draft, without revising.

6.8. If I don't have enough time to finish editing a document, I should at least check which one of the following? Choose only one and explain your choice.

The figure/table numbers and titles

The first paragraph's flow and accuracy

The last paragraph's flow and accuracy

The headings' format and wording

FURTHER READING

Chicago Manual of Style, 15th ed., 2003. University of Chicago Press: Chicago.
This is the mother ship of style manuals. This big volume contains more than you'll ever want to know about manuscript preparation, documentation, and the publishing process, but it will also have definitive answers for you on all those niggling little questions about style (do you write out the number twelve or use the Arabic numeral?). The sections on "Grammer and Usage" and "Punctuation" are a terrific reference tool for engineers. Most discipline-specific style manuals in engineering are based on the *Chicago Manual*.

Elbow, P. 1998. *Writing with Power: Techniques for Mastering the Writing Process*, 2d. ed. Oxford University Press: New York.
Chapter 3 of this influential book on the writing process describes the critical importance of "sharing" as a prerequisite for good writing.

Gladwell, M. 2001. "Wrong Turn: How the fight to make America's highways safer went off course," *The New Yorker*, June 11, 2001, pp. 50–61.
This article's discussion of how we literally misperceive reality is focused on how such misperception causes automobile accidents, but the evidence presented from research studies of perception seem very applicable to other situations as well. When we edit our own writing, we may well not see what is really there, because we have the "right" text visible in our mind's eye.

Harris, M. (2006). *Prentice Hall Reference Guide to Grammar and Usage*, 6th ed., Prentice Hall, New Jersey.

National Society of Professional Engineers. 2007. *Code of Ethics for Engineers*. http://www. nspe.org/ethics/eh1-code.asp

Samson, D.C. 1993. *Editing Technical Writing*. Oxford University Press: New York.

This book is aimed mostly at professional writers and editors and focuses on editing as a collaborative process. But three of the chapters are relevant for engineering students. The two chapters on editing text and editing graphics offer guidance and examples on editing with audience and purpose in mind. And the chapter "Degrees of Edit" gives some very practical advice for doing the most with the time you have. The reality of professional life is that you simply don't always have time to do a heavy edit. But that's no excuse for not doing at least a light edit. The examples presented are clear and useful.

Speaking: Do I Really Have to Stand Up and Talk in Front of All Those People?

By reading this chapter, you will learn the following:

- how to plan and outline your presentation;
- how to plan content and visuals for your talk;
- to design effective slides;
- to design posters;
- how to use voice quality and body language to deliver your message;
- to use the most appropriate delivery mode for your talk;
- to use your nervousness to prepare for and deliver a talk;
- how to handle questions from the audience;
- how to prepare for a team presentation.

7.1 INTRODUCTION

Engineers have to communicate information orally all the time—sometimes very informally through one-on-one conversations, sometimes a little more formally through project briefings in a conference room, and sometimes much more formally as a speaker in a big auditorium. In all these situations, you want to get comfortable with using body language, voice quality, and visuals to your advantage. You can use those tools to help deliver the message of your talk, whether you are making recommendations to a client, requesting funds for a new project, or giving the results of a research study. It is all about interacting with your audience. If the members of your audience (whether composed of 1 or 100 people) feel you are speaking directly to them and would be willing to *listen* to them in return, you will have succeeded. After all, your audience probably prefers hearing and watching your presentation than having to read a long report on the subject.

KEY IDEA

Most people would rather listen to someone explain technical information than read about those same concepts. Your audience is on your side.

Of course, most people are nervous about speaking in public, especially in a professional situation. But controlled nervousness can be used to your advantage, to give you energy and motivation. In this chapter, we will look at techniques for controlling nervousness. The most important strategy is to be prepared. There is no substitute for planning and rehearsing your talk in advance.

7.2 PLANNING THE CONTENT OF A PRESENTATION: OUTLINING

The first step in planning a good presentation is usually to make an outline of the contents. The next steps in developing your presentation would be to create the visuals and then rehearse. These steps are not always linear (you might, for instance, start by preparing a graph of test results), but you want to allow time to accomplish all three steps.

Here is a plan for step one, outlining your talk. Consider these items as you outline:

- Know your **audience**.
- Choose **three to five main points** to **highlight**.
- Divide your talk into **three sections:**
  - **Repeat** the main points in each section.
  - Decide where **visuals** will be useful.

As the talks you give become more complex, you may find that you need to make more than three to five main points, but this rule of thumb is useful to keep in mind throughout your career, especially when talking to a less technical audience.

7.2.1 Audience

You never know who will come to a presentation you are giving. Even if the audience is supposed to be just your boss and a couple of other project managers, your boss might rush in at the last minute and tell you he or she is bringing a couple of visitors from corporate headquarters to your talk. Or it might turn out that a couple of inspectors from the regulatory agency want to hear what you have to say about the air-pollution studies you have been conducting. Even at technical conferences, you cannot be sure these days that everyone has the same technical background you do. Interdisciplinary projects and studies are becoming increasingly common, and you have to be sure that everyone understands your information. Sometimes, this diversity of audience means that you must add more background information and explanation than you might think is necessary. Remember, too, that we all hear differently then we read. When we read, we can go back and reread information we do not quite understand. Your listeners cannot do that, so you have to enable them to understand your presentation from the beginning.

7.2.2 Highlighting

KEY IDEA: You only have time to present and support three to five main points!

When you are presenting, time tends go to by faster than you thought it would. You have to build in time for pauses and (sometimes) more lengthy explanations than you had originally thought. So, do not try to cover too much material! You cannot begin to cover all of the material that is in your report, thesis, or proposal. So choose three to five main points to highlight, and build your talk around them.

7.2.3 Three Sections

That old chestnut about an essay having three main parts—beginning, middle, and end—is really true for a talk. Listeners need a lot of structure to keep their mental place in listening to a talk. Studies have shown that even the most well-intentioned listeners retain less than one-third of the information given in a talk. If you think about how *you* listen, you will realize that you fade in and out of attentiveness and do not even hear everything that is said. Also, you may not understand all of a speaker's information or be able to put it in a meaningful context. Help *your* listeners remember the important points of your talk by repeating the points throughout the presentation. After all, if you state a critical point three times, you are pretty likely to get close to a 100% chance that the listeners will remember it ($3 \times \frac{1}{3} = 1$, which you can think of as 100%).

While you are making your outline, you can also be thinking about where you will need to show visuals and which of these visuals should convey technical information and which should keep listeners on track by indicating what sort of information the talk is about to cover (conclusions, recommendations, etc.).

Plan for a 15-minute talk and you cannot go wrong. Most individual student presentations cannot be longer, and most individual presentations at conferences or meetings are a similar length. At conferences, there will probably be a session moderator keeping track of time, and your talk most likely will be sandwiched in with at least two others during a 60-minute session. Allowing for questions after each talk, this situation means 15 minutes maximum for your talk itself. If you are told that you have more time than that, terrific. If you planned for only 15 minutes, then you can always add a few more details. (Somehow, there is usually more to say about a subject!)

Here is a checklist for the first part of a typical 15-minute presentation:

Introduction (2–3 minutes):

- Introduce yourself.
- Motivate the audience to listen to you.
- Explain the purpose of your work and of the presentation.
- Preview the main ideas.
- Establish key concepts.
- Do not give detailed background right away.

Beginnings are important. The impression you initially make will stay with your listeners. Begin in an upbeat manner; remember that you are meeting and greeting this audience. Try to use your voice and movements to convey some excitement about the project you are about to discuss. If you do not sound excited about the work you have done, why would an audience want to listen to you for upwards of a quarter of an hour? In your introduction, preview the main ideas you will discuss and mention any key concepts you want folks to keep in mind while they listen to you. One example of such a key concept might be to remind them that your test results are preliminary and need to be confirmed in the next phase by field studies. Do not give a detailed background of the problem or the research you are going to discuss until you have introduced these key ideas and concepts. Start with the big picture. Do not forget to introduce yourself if no one does it for you: Who are you, where are you from, and why are you talking to these people?

You do not want to memorize your talk word for word (because you would sound too stilted), but you might want to memorize the first few lines of your own introduction. You will probably be pretty nervous, and if you can begin smoothly, you will become more confident as you go along.

The following is a checklist for the second part of the presentation:

Body (8–10 minutes):

- Use an easy-to-follow organization.
- Use transitions.
- Provide supporting data.
- Bring in key concepts.
- Separate background information from your analysis.

Keep the body of your talk well organized with good graphics, and you will probably keep people awake. That is no mean feat in this age of overwork and multitasking! Be sure to discuss your topics and key concepts in the same order that you introduced them in the introduction. Keeping the same sequence is a mnemonic device that helps your listeners remember the information. Especially in the body of your talk, where you present a good number of details to support your main findings or points, you must focus on logical, repetitive organization and repetition of keywords. Be sure to keep background information separate from your *analysis*. Background is what you found out; analysis is what you make of what you found out. Your analysis is your contribution to solving the problem.

Here is a checklist for the third part of the presentation:

Conclusions (2–3 minutes):

- Provide a clear wrap-up.
- Comment on the significance of the results.
- Do not just stop in your tracks!
- Ask for questions.

Never just stop and say, "That's all." Plan a clear ending to your talk that emphasizes the most important fact or conclusion of your work. Remember that many in your audience will probably remember best the last thing you said (other than, "Are there any questions?"). Especially if your talk has been fairly technical, you must use the ending to underline the significance of what you did, found, or concluded. For talks that give instructions of some kind, finish with the most important fact, whether it is a safety warning or a statement of how easy these instructions will make the job or how important the job is. Repetition is your friend here. Do not fail to say something just because you have already said it.

Of course, you always end any talk with a request for questions or comments from your listeners. Your job is to make sure your message came through, and fortunately, presentation allows you to know for sure whether you succeeded. Asking for questions gives you another chance to be clear if anyone is confused or in doubt about something you have said or done.

On the following page is a sample outline for a 15–20-minute talk given to the U.S. Environmental Protection Agency. The talk was a progress report on a research project testing a new material for possible use as a landfill liner. You can use this outline as a guide while preparing a talk of your own. The wording you use in your outline may find its way onto your slides, but keep in mind that *these outline sections are not yet slides*! The outline is on the following page.

7.3 PLANNING THE VISUALS

When we get information in both words and images, we understand better. A picture may be worth a thousand words, but we still need words to give us the context for the picture. Be sure that your words and your visuals are in synch all the time: Do not leave a visual up while you talk about something else. Also, it is OK for the presentation screen to be dark or blank sometimes. Remember that *you* are giving the talk, not your fancy pictures.

Do not try to use too many visuals; it is better to explain a few things thoroughly than to have to skim through and skip visuals because you are running out of time. You have to allow time for pauses and for potentially more-thorough explanations than you had anticipated because you can see the audience is confused.

Sample Presentation Outline

INTRODUCTION

Project: Designing and conducting bench scale hydraulic conductivity tests of manufactured bentonitic blanket materials for landfill liners and caps.

Purpose of Project: To provide data to EPA on whether these new materials are as effective in preventing leachate from escaping into soil (and rain from getting into landfill) as are current liners: 3 feet of compacted clay.

Purpose of Talk: To update audience on testing and results so far. To solicit feedback on test design and methods and on preliminary conclusions.

List of Main Topics (What I will discuss)
     Background on material
     Testing equipment
     Testing procedure
     Current status of testing

List of Key Concepts
     Designers in a hurry to use new material
     Test designed only for ideal lab conditions
     Further testing needs to be done

BODY

Main Points

1. Background on liner material
- composition, cost, ease of installation

2. Testing equipment
- 9 steel tanks, gravel, water, etc.
- collection dish under drain

3. Testing procedure and duration
- 3 sets of tests for each tank
- each set takes two months
- readings taken weekly
- hydraulic flow calculated: $Q = kiA$
- test stopped when steady flow is reached

4. Current status
- first set of tests complete
- hydraulic conductivity values calculated
- second set of tests set up

CONCLUSION

Tentative Conclusions
     1. Conductivity is within range published by manufacturers of 3 brands tested.
     2. Conductivity is comparable to (actually slightly lower than) clay liner.
     3. Further testing is needed: e.g., simulating actual landfill conditions.

KEY IDEA

Planning Visuals

- Begin planning visuals as soon as the outline of your presentation is completed.
- Plan on using no more than 10–15 visuals for a 10–15-minute talk (at least one minute/slide).
- Consider all delivery methods:
  - Computer projection is the most professional.
  - There may be reasons to consider other methods as well.

7.3.1 Modes of Delivery

Computer projection has become the professional standard for delivery of most presentations. There are tremendous advantages and a few potential problems with using computer projection. If you have designed your slides using presentation software, you can update them even at the very last minute. There is also no chance of getting the slides out of order (as with transparencies or 35-mm slides). You can also use the full range of multimedia options available to you through your computer and the projection capabilities at the presentation site. To gain these advantages, however, you must understand and be confident about the equipment you will be using. When problems arise, it is generally because you have not checked out the equipment beforehand, especially when you are off site, away from school or the office. It is possible that the projector will not be compatible with your computer or that the right cables will not be at hand. If you have your presentation on disk, the borrowed computer may not have the right software to run your files. Or, if some of your presentation depends on the Internet or other network connections, the correct hookups may not be available. If at all possible, you should rehearse or at least visit with the equipment to be used. Go there the day before and assess the equipment.

At this point in the early 21st century, the most commonly used presentation software is Microsoft PowerPoint. This package is very user friendly and very powerful. I have seen students learn to create decent slides within 10 to 15 minutes and then learn the finer points of designing with PowerPoint within the next hour. All you have to do is open the program and start experimenting. Use the templates to give yourself ideas on design and format, but remember that others may be using those very same templates. Your talk should be unique, because both you and the way you present your particular information are unique. PowerPoint is very customizable, so take the trouble to experiment with design and format.

Suggestions for Designing with Presentation Software

- Use the design templates rather than the genre templates ("sales pitch," etc.). Design templates offer a consistent format for slides (font size and type, placement of titles, etc.), but do not try to dictate content.
- Be careful about colors: Use very dark text on a very light background or vice versa.
- Avoid red text, as it is difficult to read from a distance, no matter how large the text.
- Either keep slides consistent in design or vary the design according to subject matter. For example, if you are introducing a multiphase project, you might make the introduction slides slightly different in format from the body slides. Conclusion slides might then repeat the introduction slides' design.
- Experiment with animations and dimming, but consider your audience before using these effects: A semitechnical audience may appreciate "entertainment" more than would the members of your dissertation committee.

Packages such as PowerPoint help convey information in manageable chunks. This sort of software has revolutionized business and technical presentations by forcing the presenter to distill his or her thoughts and information for greatest simplicity and clarity. Bullet points or talking points can be created effortlessly in PowerPoint, and the product has a large selection of clip art just a click away. See how the slide shown in Figure 7.1 clearly introduces a case study on environmental contamination, using a piece of clip art to render the subject matter more memorable in the mind of the listener. Be careful, however, of overusing clip art—use it sparingly on your slides.

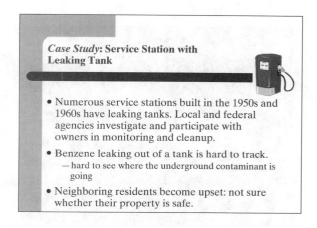

Figure 7.1
Sample slide for introducing a case study.

You can use presentation software not only to create the visuals for your talk but also to organize your thoughts (and therefore your talk) and to create handouts for the audience. Once you start using the software, you will realize how the necessity to type your thoughts in logical order, using short phrases or sentences, helps to organize your thoughts in the first place. Chances are that if you just started extemporizing (speaking off the cuff) about the difficulty of monitoring a site suspected of being contaminated with benzene, you would say a lot more than what's contained in the second bullet point on the slide in Figure 7.1. Yet, for a general audience, that bullet point contains a good summary of the problem.

Be careful, however, of letting your bullet points speak for you. Those bullets are simply the tip of the iceberg; it is up to you to provide the connections between the bullet points that make them come alive. You want to create *meaning*, not simply a *list*, out of your data and your analysis. *Never stand in front of your audience and read your slides.* Use the slides for your own visual reference, and repeat keywords, phrases, and critical pieces of data, but always add more information, further examples, or a story that illustrates the point indicated by the bullet.

Do Not Let Your Bullet Points Speak for You

- Your job is to create meaning so the audience understands the significance of your work.
- Do not simply read your bullet points.
- Explain: Provide "connective tissue" between bullet points.

KEY IDEA

Never let your visuals upstage you or the content you are presenting. A package such as PowerPoint supports all sorts of multimedia applications (including animation, video, audio, and other special effects) that can be used to make your talk understandable, memorable, and entertaining. But as you plan your visuals, always consider first the audience to whom you will be presenting. The more technical knowledge the audience members have about your subject matter, the fewer bells and whistles they will need. And remember that bells and whistles should never be used for their own sake, but always in the interest of stressing or explaining a particular point that you are making. For example, the text on the PowerPoint slide shown in Figure 7.2 does not match the illustration.

Figure 7.2
Example of a possibly frivolous slide.
[If this slide is not meant to illustrate something about cats or catlike qualities, then find another picture. Never use a picture just because it is "cool" or you like it.]

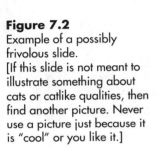

There are situations in which computer projection may be unavailable or may not be the best choice for your talk. If you are showing photographs that were not computer generated, you may want to use 35-millimeter slides to achieve the best resolution. If you are leading an interactive meeting or workshop, you may want to use overheads (transparencies) to capture the ideas or modifications suggested by participants. Sometimes, the overhead projector will be the only equipment available to you.

Whatever your mode of delivery, keep these guidelines in mind as you develop your visuals: Make sure that each visual makes the right point. Do not get carried away with the possibilities offered by clip art and by all the many pictures you can download from the Web. Let your talk's content guide you, and then try to build in a little gentle humor here and there.

7.3.2 Designing Posters

An increasingly common way to present information is through a poster display. The audience members can walk around among the posters, stopping at those that interest them and talking to the poster author. This setup means that you or another of the poster's authors stands beside the poster. In order for people to decide whether or not to stop and talk with you, they need to be able to see the poster (at least the title, some of the graphics, and the headings) from across the room (or at least 15 or so feet away). Your poster can make more use of sentences than your slides, but you still must be sensitive to the needs of observers to read the poster quickly.

When considering the content for your poster, think in the same way about the audience as you would for a presentation; What do they already know? The typical poster board is $3' \times 4'$ – that may seem like a lot of room, but not when people have to read on the fly. So, don't provide much background; focus on the analysis and the results of your work.

Here are some suggestions for organizing and designing the information on your poster.

- Organize posters for clear sequence and readability:
    Use grid as frame
    Start upper left; finish lower right.
    Make vertical (preferable) or horizontal alignment *clear*.
    Provide *lots* of visuals with your text.
    Use bulleted lists, not just paragraphs.
- Provide visually contrasting background.
- Make sure the font is big enough:

    At least 54-point font for title
    At least 36-point for headings
    At least 24-point for text (axes might be 18-point)

Test your poster on friends or classmates. Ask them so stand at least 15 feet away and see whether they can read the headings. Listen to the questions they ask about the content, and try to incorporate your answers into the poster itself.

Guidelines for Designing Visuals in All Modes

- Design each visual to make one main point.
- Two visuals may be better than one.
- Be sure each visual makes the right point.
- Leave a visual up only as long as you are talking about it.
- Remember that *you* must still be the focus, not your visuals.
- Avoid clutter. On graphs, be sure to omit needless gridlines, tick marks, etc.
- Check your visuals for distortion, spelling errors, and computational mistakes.
- Think big.
- Create meaningful titles.

7.3.3 Designing Slides

Here is a checklist for designing slides:

- *Simplify* any hard-copy graphics (graphs, tables, etc.).
- Think big.
- Include mapping visuals.
- Put main point at top of slide.

The following subsections explain these four points.

Simplify

A long table or a complex graph that works in a report will *not* work well in a presentation. The audience does not have enough time to process and understand visuals that are very complex and dense. Remember that each visual will probably be shown for less than one minute. Consider Table 7.1.

Table 7.1 does not make a good presentation graphic. There are too many numbers to absorb and understand quickly, and the significance of those numbers is

Table 7.1 Comparison of Settling for Various Types of Piles.
The piles that exceed 0.25 inches of settlement are shown in red. The piles that are shaded in gray were discarded. [*This table is too crowded to make a good presentation visual. Titles for presentation graphics should not have captions or numbers.*]

Concrete Pile			Hollow Steel Pile			Concrete-Filled Pile		
Diam (in)	Len (ft)	Sett (in)	Diam (in)	Len (ft)	Sett (in)	Diam (in)	Len (ft)	Sett (in)
12	50	0.19	12	50	0.07	12	50	0.19
12	55	0.37	12	55	0.18	12	55	0.37
12	60	0.36	12	60	0.17	12	60	0.36
12	65	0.61	12	65	0.28	12	65	0.61
14	50	0.20	14	50	0.10	14	50	0.20
14	55	0.33	14	55	0.16	14	55	0.33
14	60	0.43	14	60	0.21	14	60	0.43
14	65	0.68	14	65	0.26	14	65	0.58
16	50	0.23	16	50	0.10	16	50	0.23
16	55	0.32	16	55	0.20	16	55	0.32
16	60	0.41	16	60	0.42	16	60	0.41
16	65	0.43	16	65	0.43	16	65	0.43

not made instantly visually clear. The numbers showing diameter, for example, look as important as the numbers showing settlement (presumably the critical information). But the main problem with this visual is the small-sized and wordy caption; the labeling is left over from the hard-copy version. Presentation visuals do not need to be numbered, and the titles should be very simple and short, usually without captions. In any case, the caption here is meaningless in the hard-copy, handout version of your slides.

For visuals that are mostly words, remember again that an audience does not have time to read and process information the way a reader of a book can. You must use fewer words and allow more white space between the words. Here is a general rule of thumb for text on slides:

- No more than seven or eight lines of text per slide.
- No more than seven or eight words per line.

Think Big

One of the prime benefits of using a software presentation-package such as Microsoft PowerPoint is that the default text sizes are always large enough to be seen easily. Because we are so used to judging visual ease of reading from a small screen or from a hard copy, we forget that projecting text and images over distance necessitates a larger scale if the visuals are going to be read easily. If you customize the default font settings for any particular design template in PowerPoint, make sure that you maintain a font size that can be read easily from the back of a large room. If you are using an overhead projector, always enlarge the hard-copy versions of your tables and graphs. Use these minimum font sizes for presentation graphics:

- 18-point minimum font for labeling axes
- 32-point minimum font for titles
- 24-point minimum font for first-level bullet points
- 20-point minimum font for bullet points and main text (only for second- or third-level bullet points)

Use fonts larger than these wherever you can. A good test for readability of your slides is to print them out in slide view, place each one on the floor, and stand back about a foot. If you can read the slide easily, then it is probably large enough for your audience to see easily when it is projected on a screen in front of a large room.

Include Mapping Slides

You are probably aware of the importance of content visuals in conveying information on the subject you are discussing. For example, the graph in Figure 7.3 plots the theoretical speed and position of a ball. The visual shows the exact relationship faster than words alone could explain it.

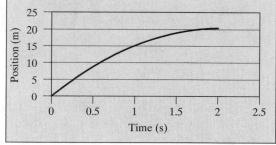

KEY IDEA: **Content-oriented slides** display graphs, bar charts, etc.—mostly images.

Figure 7.3
Example of content-oriented slide for a presentation.

You may not be aware, however, of the importance of visuals that map your talk and emphasize certain key points (e.g., how the theoretical information was used to solve a problem). You want to tell your audience where you are in the talk and what to expect. A mapping slide that outlines the stages of the research project being reported, as shown in Figure 7.4(a), prepares the listener for what is to come, thereby enhancing his or her understanding of ideas and concepts as you explain them in more detail. The slide in Figure 7.4(a) would come toward the beginning of the talk and would communicate several key ideas the listener:

- The experiment is not original.
- The researcher is interested in repeatability of the experiment.

Figure 7.4(b) would come later in the talk as an introduction to analysis of applicability of the water-treatment method studied. The goal of the research project would have been made clear in the beginning of the talk, and this slide refers back to that goal. Note that the author of the photograph is given credit right on the slide, in smaller letters under the photo. You always want to credit the sources of visuals that you did not create yourself.

The experimental phase of the project had two stages.

1. Duplicate Browning's experiments using zinc to treat drinking water.
2. Analyze our data and interpret any differences from Browning's results.

(a)

The third phase of the project addressed the goal of the experiments.

- Determine applicability of the zinc method to drinking water-treatment systems in northeastern Ghana.

Photo: Katherine
Alfredo 2005

(b)

Figure 7.4 (a,b)
Example of two mapping slides for a presentation. You should use both analytical and mapping slides.

KEY IDEA: **Mapping slides** map the structure of the talk—mostly words.

The first mapping slide should always be your title slide. Tell the members of your audience both orally and visually who you are, where you are from, why you are talking to them, and what is the general subject of your talk. In Figure 7.5, the font sizes are Arial 40 for the title, Arial 32 for the presenter's name, and Times 24 for the speaker's affiliation and date. Arial is a sans-serif font (meaning that it has no little "feet" at the edges of the letters), so it projects very well. Times is a serif font that should be used carefully for presentation graphics. Use Times for information that is somewhat less important than other information, or else make sure you never use less than a 24-point size.

Many instructors feel that an outline-of-the-talk slide should follow the title slide. Indeed, you want to orient the reader to the main ideas you will be discussing, but be

Figure 7.5
Sample title slide for a presentation.

careful of making your outline either too generic or too wordy. How is your audience supposed to remember 6 or 7 points that your talk will cover? See Figure 7.6 for an example of a generic outline slide that conveys no specific meaning about the specific talk to which it refers. Even if the speaker fills in the blanks — What methodology? What research? — the slide itself is useless. If you are going to make an outline slide, Figure 7.7 is a better example. One mapping slide that always seems to make sense is "Conclusions." Labeling that slide and writing the conclusions in simplified form tells the listener that the talk is almost at an end. If his or her mind has been drifting, now is the time to listen attentively and get the talk's most important points.

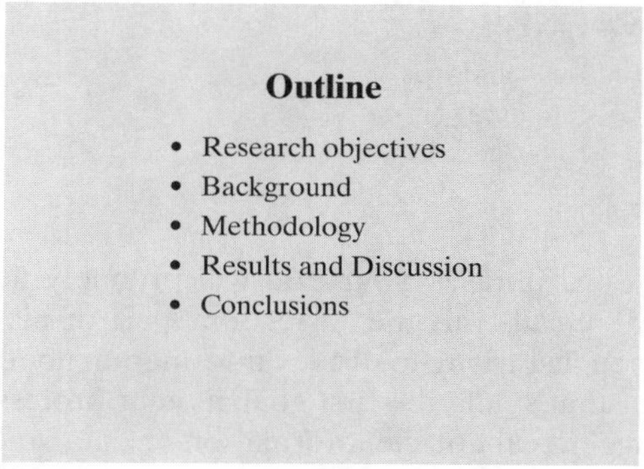

Outline

- Research objectives
- Background
- Methodology
- Results and Discussion
- Conclusions

Figure 7.6
Example of a generic – and therefore useless – outline slide.

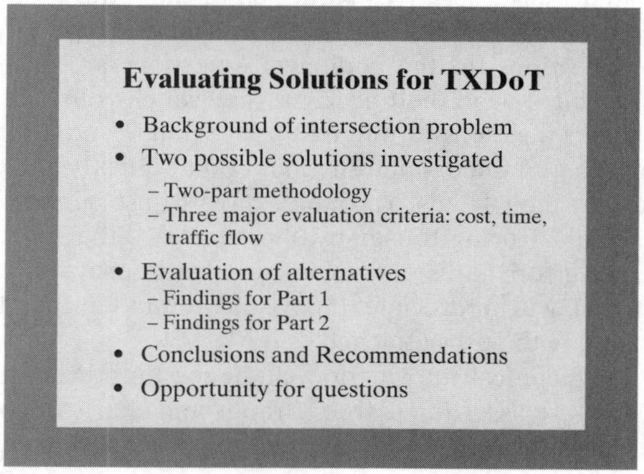

Evaluating Solutions for TXDoT

- Background of intersection problem
- Two possible solutions investigated
  - Two-part methodology
  - Three major evaluation criteria: cost, time, traffic flow
- Evaluation of alternatives
  - Findings for Part 1
  - Findings for Part 2
- Conclusions and Recommendations
- Opportunity for questions

Figure 7.7
Example of a more meaningful outline slide.

Put Main Point at Top of Slide

Along with the convenience and effectiveness of a presentation-software package such as PowerPoint comes the danger of abusing or misusing its power to organize information into hierarchical bulleted items. A slide like the one in Figure 7.8 cannot be understood out of context and leaves the audience with glazed eyes from trying to connect together too many bulleted items.

Several writers have recently published articles about the danger of using PowerPoint to oversimplify information or just to bore the audience to death (see Further Reading). The most compelling of these arguments has been put forth by Michael Alley, who suggests a pattern for slides that places the main point of the slide in the position of title. The main point of Figure 7.8 is not "planning," but rather something like, "Consider these factors when planning a talk." Indeed, once you create a title-sentence, you realize that the bulleted list is mixing apples and oranges: the first five items are cautions or concepts and the last five are instructions. This slide should probably be two slides. Since most main points worth discussing

Planning

- Different retention rate
  - Listening vs. reading
  - Talking vs. writing
- Listeners not in control
- Audience not all technical
- Presentation linear
  - Thinking not linear
- Easy to get off target
- Use an outline
- Divide into 3 parts
- Build in visuals
- Introduce yourself
- Don't stop in tracks at end

Figure 7.8
Slide with too many bullet points without any context and without a meaningful title.

are not one or two-word phrases, your title will probably need to be a sentence much of the time. Alley calls this the "assertion" part of his design scheme. Why "assertion"? Aren't you just trying to convey information about something you did? Well, yes, you are, but that's only the first goal: in your professional life you will be paid for what you conclude about the information or data you have gathered. How do you interpret the data? What plan or process do you think is best? How well is the project keeping to schedule? To answer any of these questions you must assert something. And most of your slides, no matter what the type of presentation, are meant to answer questions for the audience. Assertions are answers to questions that your audience will have in their heads as they watch your talk.

Figure 7.9 shows a slide with minimal bullet points and a sentence for a title. The audience can see and understand the main point quickly and can keep the two bulleted sentences in their heads. You don't have to use sentences as your bullet points, but be careful of using too many short phrases that have no context and therefore no meaning for the listener. You can always provide the context as you speak, but even you will appreciate the visual reminder that these bullets are instructions (actions) with verbs attached.

Especially for technical information, audiences appreciate seeing visual evidence that what you are asserting is true. Graphs and charts of course provide this evidence, but so might photographs or a picture of something that will help the reader remember the point you are making. Figure 7.10, the photograph I took in New Orleans several months after Hurricane Katrina, is compelling visual evidence of how effectively visuals make your point. Of course, the visual need not take up so

Short sentences make meaningful titles.

- Consider using title font smaller than the default.
- Try to keep bulleted lists to a minimum.

Figure 7.9
Example of a slide using a sentence (an assertion) for a title.

Make as much information as possible visual.

Figure 7.10
Example of a slide using the assertion-evidence design scheme

much room on the slide, nor be so dramatic. But some visual evidence or reminder of what you are asserting is helpful to your listeners.

Using the main point as the title of many of your slides will definitely mean that you revise your slides several times. Most of us tend to make topic outlines for our talks — outlines of the subjects we will cover. But what the audience is interested in is what we make of these subjects — our assertions. Once you take the time to revise your slides with more meaningful titles, however, your reward is that you will never go blank or be stumped when you transition to a new slide. Your main idea — what you most want to get across to your audience — is right there at the top of your slide.

PRACTICE QUESTION: DESIGNING SLIDES

PONDER THIS:

How could the slide shown in Figure 7.11 be improved? Identify specific problems with it.

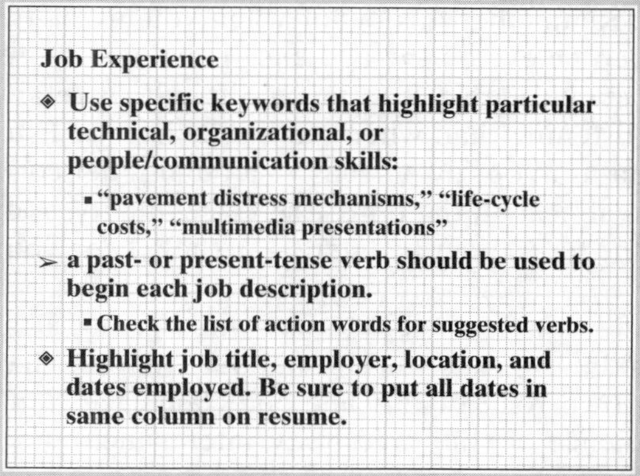

> **Job Experience**
> ◆ **Use specific keywords that highlight particular technical, organizational, or people/communication skills:**
>> ▪ **"pavement distress mechanisms," "life-cycle costs," "multimedia presentations"**
> ➢ **a past- or present-tense verb should be used to begin each job description.**
>> ▪ **Check the list of action words for suggested verbs.**
> ◆ **Highlight job title, employer, location, and dates employed. Be sure to put all dates in same column on resume.**

Figure 7.11
Example of badly designed slide.

Suggested Answer

Some of the problems with the slide include the following:

- Too much text.
- Awkward placement of text.
- Inappropriate background for a text-laden slide—the lines are distracting.
- The size of text doesn't always reflect the level of bullet.

- The bullet design is not consistent.
- Capitalization is not consistent.
- Meaning of title is not immediately clear.

Figure 7.12 shows a suggested revision.

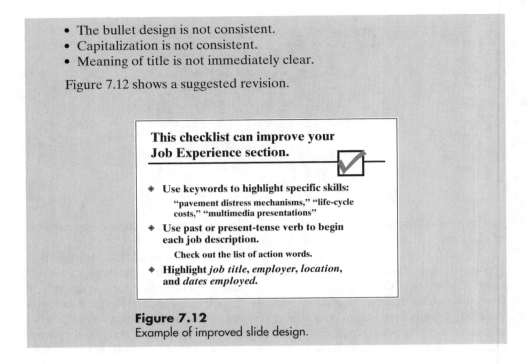

This checklist can improve your Job Experience section.

- Use keywords to highlight specific skills:
  "pavement distress mechanisms," "life-cycle costs," "multimedia presentations"
- Use past or present-tense verb to begin each job description.
  Check out the list of action words.
- Highlight *job title*, *employer*, *location*, and *dates employed*.

Figure 7.12
Example of improved slide design.

7.4 PLANNING THE DELIVERY

Remember that your audience will need your help to understand clearly what you are saying. After all, readers can read at their own pace and stop whenever they want to. Listeners, on the other hand, are at the mercy of the speaker's pace and level of detail. Remember this factor when you give a talk. Plan not to talk too quickly, and keep an eye on the audience members to see if they are understanding you. For many people, it is more difficult to retain large amounts of information when listening than when reading. When we listen to a talk, we are also listening to what is in our own heads — all that inner conversation that most of us have going on almost all the time. Use voice quality, body language, visuals, and transitions — the tools described here — to help keep your listeners with you during your talk.

KEY IDEA: Enthusiasm for what you are talking about is the most important ingredient of a successful talk.

You have some powerful tools to use when giving a talk. Aside from the words you speak and the visuals you show about the topics indicated in your outline, you have your own voice quality and body language. You can use these tools to convince an audience of your credibility, intelligence, and enthusiasm for the work you have done. Enthusiasm may be the most important ingredient of your talk: If you show that you care intensely about the subject and the work, chances are your audience will also care. Conversely, if you speak in a monotonous, flat tone and if you stand in one place or slouch, the audience members will get the message that you do not care that much, so *neither will they*. The best way to improve your delivery strategies is to harness the enthusiasm that should already be there inside you (though perhaps rather buried). Positive energy communicates best. But first you need to get feedback on how you come across to an audience.

You can use rehearsing, videorecording, and anticipating questions as strategies for eliciting feedback.

7.4.1 Rehearsing

The best way to improve the delivery of your talk is to rehearse in front of someone. Rehearsing in front of at least one other person enables you to get constructive feedback—on what you are doing right and what you are doing wrong—that will help you communicate effectively with your audience. Ask your classmates or coworkers to watch you run through your talk, and offer to do the same for them. Rehearsing at least three times is a good idea for less experienced presenters: Rehearse once on your own, once standing up in front of someone, and once (ideally) with your audiovisual equipment in the space where you will be giving your talk. Ask your surrogate audience members to critique your delivery. Could they hear and understand everything you said? Could they follow the talk, and did you make it interesting? Here is a suggested checklist of questions for your reviewers to answer. You can also use these questions to evaluate your own talk when viewing your video.

Questions for Your Rehearsal Reviewer to Answer about You

Is the structure of the talk evident?
Do the various sections hang together and form a whole?
Does the pace seem right?
Is the subject matter clear and compelling?
Are the visuals clear and compelling?
Does the presenter sound enthusiastic?
Does the presenter speak to the point without using verbal fillers, such as "um" and "you know"?
Does the speaker maintain eye contact?
Does the speaker move naturally, without fidgeting?
Could I hear every word of the presentation?
Does the talk hold my interest?

Voice Quality

Tone of voice is so very important in any interpersonal communication. Think of how many times you may have been misunderstood because your listener thought you were being sarcastic or bored when you really were not—your voice just somehow sounded that way. A good voice is an enormous personal asset. Your career will truly benefit if you can learn to control, modulate, and use your voice to be persuasive, informative, or inspiring, as the occasion demands. In a presentation, you have to ensure first of all that you are speaking loudly enough for everyone in the room to hear you—everyone, even those people in the very back rows. Every speaker can learn to speak more loudly. Try the exercise explained in the upcoming box to increase the volume of your voice. The trick is not to put more pressure on your throat—that is not where your voice originates—but rather to apply pressure to your diaphragm, where your voice originates. Your diaphragm is like a bellows that can heat up your voice to a roaring fire.

Practice Exercise: Increase Voice Volume and Resonance

- Stand with your feet shoulder width apart and your hands on your diaphragm, between your belly button and breastbone. Open your mouth and emit a constant sound at the same pitch. Press on your diaphragm and see how your voice volume increases without your even trying. The harder you press, the louder is the sound.
- Remember that your voice does not come from your throat; it rises up from your diaphragm and gains power from the air in your lungs.

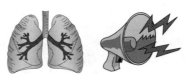

Second, you want to work on the richness, pleasantness, and variety of your voice. These qualities are controlled by what sound experts call *tone*, *pitch*, and *dynamics*. Some of us naturally have a rather monotonous voice, one that stays at the same pitch, tone, and dynamic. In any talk, we want, however, to use our voice as an instrument that conveys emphasis, importance, humor, and sometimes even emotion. *Tone* is the resonance of your voice. If you have a squeaky or thin voice, check out the voice exercise in the previous box for a method of strengthening your voice so it sounds richer and more resonant. *Pitch* is the highness or lowness of your voice. Try to vary your pitch to emphasize important points. Remember that your audience understands meaning by the way that your voice rises and falls. A rising voice at the end of a sentence is the way that we understand you are asking a question. Do not make or let your voice rise at the end if you are not asking a question. However, do not let your voice fall so low at the end of a sentence that the listener cannot catch the last few words. Be careful not to run out of energy (and therefore volume) at the end of your sentences. *Dynamics* refers to loudness and softness of a voice. Again, you can emphasize important points by raising your voice volume a little.

Also be careful not to speak too fast. Nervousness makes us speed up. On the other hand, if the pace of your talk is slow, it may seem that you are unsure of what you are going to say next. Strive for a medium pace, one that carries listeners along without losing any information because you are whizzing over the words.

KEY IDEA

To Improve Voice Quality:

- Speak loudly enough for all to hear.
- Make sure your voice is varied in pitch and dynamics.
- Speak in a resonant tone.
- Make your voice rise and fall appropriately; don't run out of steam at the end of a sentence. But don't let your voice rise if you are not asking a question.
- Speak at a medium pace. Generally, speak more carefully and slowly than you normally do.
- Speak *toward* the audience. Glance toward slides only briefly.

Body Language

You want to show that you care by leaning slightly forward, toward the audience, and using your hands appropriately to emphasize points and to include the audience in your thoughts. If you stand with your arms crossed, lean away from the audience, or slouch against the table or podium, your gestures are excluding the audience and creating a barrier between you and those to whom you are speaking. Any audience member wants to feel as though you are speaking directly to him or her. If you stand in one place all the time, your audience will become bored, and chances are you will become increasingly nervous. Moving around helps control nervousness, because movement relieves stress. Be careful as you move around, however: Move naturally, and do not develop annoying repetitive movements without realizing it. I once watched a young speaker drive her audience crazy by continually scratching behind her left knee. She had no idea she was doing this, but the audience was mesmerized by this repeated nervous motion and consequently could not pay attention to the admirable content of her talk.

The other way to distance yourself from and alienate an audience is by failing to make eye contact. In the United States, if someone does not look at us, we either lose interest in what he or she is saying or feel that the speaker is untrustworthy. The same reaction may not occur in other cultures however; in fact, if you are from a non-Western country, you may be uncomfortable looking directly at your audience. Nevertheless, if you are speaking in the United States, you should know that looking someone in the eye is considered to be a trait of truth and strength. If you gaze over the heads of your audience, down at the floor, or, worse, out the window, your audience will probably simply stop listening to you. If you are not looking at your audience because you are afraid of what you will see, relax. Most listeners are glad to be hearing your information delivered in an accessible and visually appealing way. They are on your side, and if you look out at that sea of faces, you will probably see a lot of friendly, positive expressions. You can get a nice charge of energy by looking at your listeners and seeing someone nod as though he or she understands or even smile as though he or she is happy to have this information. Do not cut yourself off from this positive energy by failing to look out at your audience. On the other hand, do not stare at just one or a couple of people. Practice a sweeping motion, resting your eyes for just a couple of seconds on each face. Try to include the entire room in this sweeping motion.

KEY IDEA

Body Language

- Use appropriate movement. Move and position your hands naturally.
- Use inclusive gestures (no crossed arms).
- Maintain good posture (no leaning against the podium or shoving both hands in pockets).
- Do not make distracting tics or noises.
- Maintain good eye contact.

Stand and face the audience!

Using Visuals

It is easy to use too many visuals, now that they are relatively easy to create. Check yourself during rehearsal by timing your talk *while using* visuals. A good rule of thumb for the number of visuals is to use no more than 1 per minute, or 10–20 for a 15–20-minute talk. Be careful of a natural tendency to include more visuals than you really need to make your points. Ask a reviewer whether your visuals support your points well, raise further questions, or seem rather empty of content. Practice displaying the visuals using whatever technology you have chosen; the last thing you want to do is to have to fiddle with the equipment during the presentation. Learn how to use a laser pointer so that you do not flash it around needlessly.

Using Transitions

Moving a reader along from point to point is difficult to do in writing; in a presentation, you have more tools—pitch and tone of voice, gestures, etc.—to help. And you have another tool that you also use in writing: transitional words and phrases. These words and phrases can be more direct and informal in speech than in writing. Rhetorical questions work really well to sum up one section of information and point to the next section: "So, what did we find once we had completed the lab experiments?" "What did our studies show us about the load-bearing properties of this material?"

Remember that in speech, you will want to use active voice—e.g., "we found deformation"—much more than passive voice—e.g., "deformation was found." Active voice will help your speech be clearer, more concise, and more direct.

7.4.2 Videorecording

An excellent way to improve your presentation skills is to have someone video-record you so that you can see for yourself what your strong and weak points are. As you watch the disk later, ask yourself these questions:

- What is my greatest strength as a speaker?
- What is the weakest part of my presentation? Why?
- What specifically can I do to improve? Should I do voice exercises? Practice more? Reorganize?
- What should be the result of my improvement efforts in my next talk?

Always notice what you do right as well as what you do wrong. Some students realize, for instance, that even though they were very nervous giving the talk, the video shows a person seemingly in control, with few signs of nervousness. This is great information to have: You do not *show* your nervousness nearly as much as you *feel* it. Knowing that should give you more confidence the next time around.

KEY IDEA: Watching the video of your presentation, notice what you do RIGHT as well as what you want to improve.

When it comes to improving certain aspects of your presentation (increasing voice volume, for instance), practice strategies for planning, rehearsing, and answering questions and see what difference the strategies make the next time around. Set realistic goals for yourself. Your voice does not have to be broadcast quality, like your favorite news reporter's. It just has to be loud enough and resonant enough to get your meaning across. Listen closely to your videodisk, and pretend to be someone in an audience listening to you. Take notes on your performance. When you tape yourself again, compare the new notes with the old. *Any* improvement is a significant improvement.

7.4.3 Anticipating Questions

The question-and-answer (Q&A) session that inevitably follows almost every presentation is an important mechanism for exchanging information. The audience gets to clarify or learn more about your subject matter, and you get to discuss some details that had to be left out of your talk or to correct any misunderstandings that may have arisen. You, as well as the audience, can learn a lot during this session. Unfortunately, many speakers are afraid of Q&A and do not even want to think about it beforehand. That is unfortunate, because preparing for Q&A will give you a good level of confidence. How do you prepare?

You can ask your rehearsal reviewers questions about the content of your talk: What is not quite clear to them, or what do they want to know more about? The first type of question will lead you to reorganize or clarify some part of your talk, and the latter type will give you a heads-up on possible questions your audience will have following your actual talk. You cannot prepare information on every topic related to yours, but being aware of future directions or similar work in other areas would enable you to answer very nicely such a "tell me more" kind of question.

As you look over your outline and rehearse your talk, think to yourself about questions that might arise from your content. Jot them down, and spend a little time thinking about them. Don't spend too much time—you cannot prepare for every type of question, nor should you have to. Your job is not to know everything, just to have done your homework on the subjects you are presenting.

7.5 CONTROLLING NERVOUSNESS

Everyone gets nervous at some point before giving a talk. The trick is to use your nervousness as a source of energy without becoming overly anxious. Let your nervousness give you energy. Being nervous means at least you are not apathetic; after all, people who are bored and indifferent are going to *sound* boring. But do not let nervousness overwhelm you. Look over the upcoming strategies for coping with nervousness before and during your talk, and see which ones might work for you. Different strategies work better for different people. The goal is to keep your nervousness within reasonable limits at the same time that you harness it for energy, motivation, and the willingness to go the extra mile in preparing.

7.5.1 Before Your Talk

Deep breathing is something you can do right before your talk or anytime you feel anxious. Simply breathe in slowly on a count of, say, four, and then breathe out on double that amount, or eight counts. Counting slows your breathing down, and soon your body will feel more relaxed. When the body relaxes, the mind tends to follow suit.

Mental imaging is a technique used widely by coaches for teams and individual players to buttress their confidence. For example, a kicker might use this technique before being brought in to attempt the extra point in a football game. He would create a mental movie that shows him moving his foot back, swinging forward, and connecting with the ball at exactly the right point. He would watch mentally as the ball arcs up into the air and sails over the goal right between the goalposts to thunderous applause from the spectators. In your case, imagine yourself getting up to give a talk, beginning confidently, and speaking at exactly the right pace, with people looking pleased and affirming by nodding their heads. Your mental movie would include your finishing on time with good energy, a definite conclusion, and perhaps a smile as you ask for questions. Remember also to picture the audience—a group of people

KEY IDEA: Practice techniques to help calm your nerves before a talk: Deep breathing, mental imaging, mind games, exercising.

with glints in their eyes of understanding and even admiration for you and the work you have done. Such a mental movie builds confidence, and confidence enables you to make the movie come true.

Mind games can help divert your mind from focusing on your own fears and anxieties. Count the chairs in the room, or say people's names backward, for example. Bring a crossword puzzle to occupy you during the time before you get up to speak. Or really concentrate on the other talks and think about their content instead of on your own talk. Last-minute cramming and memorization usually does not work and ends up making you feel more nervous by the time you actually come to the podium.

KEY IDEA

> **More Tips for Controlling Nervousness before Your Talk:**
>
> - Try exercising.
> - Avoid caffeine the day of your talk.
> - Replace negative self-assessments with positive self-talk.
> - Check all equipment in advance.

7.5.2 Just Before and during Your Talk

Here is a strategy for settling yourself down right before your talk: As you come up to the podium, speak to a few folks along the way. If you know someone in the audience, be sure and say hi to them as you move to the front; if you do not know anyone, speak to the moderator or to the speaker who went before you—tell her what a good job he or she did. You could even admit to the moderator that you are nervous; chances are he or she will say something reassuring. The point is that by carrying on these small conversations, you ready yourself psychologically to behave during your talk as though you are simply continuing a conversation, only now with lots of people. Conversing with people is different than thinking of them as some sort of enemy from whom you would like to hide. This strategy should make facing the audience less intimidating and put you in the right frame of mind to relax, focus on what you want to say, stay connected with your audience, and even have fun.

Once you are at the podium, take a moment before speaking to get relaxed and adjusted to your audience. Think of the audience as a group of friends, even if you have never seen any of the people before. As you talk, keep an eye on the audience members' reactions, and slow down or ask for questions if some of them are looking puzzled. This checking takes practice; do not let the reactions of a few folks alarm or throw you. You can hardly ever make *everyone* understand *all* of your points; do not even try. There are always some people who will look bored or distracted for particular reasons. Think about that 8:00 AM class you took; you were probably not the most lively looking listener.

Remember, your audience always wants you to succeed! They will give you the benefit of the doubt because they are happy sitting there being informed, and perhaps even slightly entertained.

Practice Question: Harnessing Nervousness

In what ways is nervousness a useful quality for you as a speaker?

Suggested Answer

Nervousness gives you energy. It makes you more alert. Use your nervousness as a source of strength and enthusiasm. Nervousness encourages you to rehearse more.

7.6 HANDLING QUESTIONS

Audience questions can give you, the speaker, very valuable feedback on the direction and methodology of your work. Here are some strategies for handling questions as they come at you:

Tips for Answering Audience Questions

1. Remain standing at the front of the room following your presentation. Do not run off or start heading to the edge of the speaking area. Convey confidence and a willingness to entertain questions.

2. Be aware of your body position; the audience is watching you. Avoid playing with your clothing, hair, notes, or writing implements. If your partner is answering a question, pay respectful attention.

3. Try to select questions from different parts of the room. In most Q&A situations, it is best to try to include the entire audience by using this method. Even if a particular person does not get a question answered, he or she won't feel ignored.

4. If you are in a large room, repeat the question before you answer it, so everyone can hear it. Paraphrasing serves three crucial functions: (1) It allows everyone in the audience to hear the question; (2) it gives you time to think of your response; (3) it ensures that you understand exactly what is being asked of you.

5. Answer each question clearly and then move on. Provide an efficient and accurate response, and move to the next question. Try not to ramble or wander into a related topic.

6. Signal when you will accept one last question. To avoid the perception that you are trying to escape a particularly nasty question, warn the audience in advance that your time for questions is quickly passing. In a classroom setting, however, this role usually falls to the instructor.

7. Brainstorm possible questions in advance. The best way to prepare for a Q&A period is to think of possible questions ahead of time. You may not accurately guess what your audience will ask, but you will feel more confident. In addition, you may very well guess accurately, and then your answer will sparkle with professionalism.

8. Do not assume that tough questions are hostile questions. You may have simply gotten someone really thinking! If a question is long and complicated, or you just do not understand it, do not try to bluff your way through an answer. Admit that you had not thought of that or studied that as part of your work, and *offer to speak with the questioner privately* after the session.

9. Practice answering questions as a part of rehearsal. If you practice with a potential audience member (and you should), have that person ask you questions about the presentation. This practice will help simulate the actual presentation experience and will boost your confidence.

Remember that the Q&A session can be the most valuable and the most enjoyable part of the presentation for you. You can learn new perspectives and ideas, and you will get feedback on the direction of your work or approach to a problem. Stay upbeat and positive, even when you are confused by a long, complex, or unforeseen question. Most listeners are not trying to test your knowledge; they really want to learn from your talk. But they have their own individual agendas, and their concerns and backgrounds are often not the same as yours. So each listener is trying to fit your information into the framework of what he or she knows and works on. Sometimes the fitting process takes some work on the part of speaker and listener, but especially in science and engineering, listeners and speaker usually share a strong bond: Everyone wants to see progress in solving that particular problem or in advancing the body of knowledge about that particular field. And when someone says publicly (as will happen to you sometime) that he or she really enjoyed your presentation, you will feel *great*.

7.7 PREPARING FOR A TEAM PRESENTATION

Many presentations in engineering practice are given by teams of engineers or by multidisciplinary teams that may include engineers, architects, planners, marketing experts, salespeople, attorneys, academic researchers, regulators, and other types of professionals. In many college courses, presentations of research are also done by teams—usually, teams of students with varying backgrounds and areas of expertise. So, learning to present well involves learning how to coordinate with other speakers as well as how to plan and deliver your own material.

All of the same strategies you might use for planning, rehearsing, and controlling nervousness apply to team presentations as well as individual ones. What changes a bit is the delivery. You have to consider not only where you will stand and how you will move and run the equipment, but also where your teammates will stand and move, etc. And you have to plan what you will do when you are not the one talking. At many presentation venues, the entire team will either sit or stand at the front of the room, so you will need to practice keeping your hands calm and trying not to fidget when you are not talking. Because a team presentation has to be even more "choreographed" than a solo presentation, rehearsing with the rest of your team, probably several times, is even more important for success.

What then, are some special strategies for preparing a team presentation? First of all, you must be sure to share contact and schedule information with each other at the beginning of the project. Meet as often as you can to plan and rehearse the talk. Second, divide up the responsibilities and the work in such a way that draws on each member's particular strengths. Each of you will most likely be responsible for a particular part of the presentation's content, but it might be best to have one person creating the slides, so that they are uniform in design. If you have the choice, you will want to select the person with the most outgoing, upbeat personality to introduce the talk (and possibly also to conclude it). It may be best to have this person act as a sort of emcee, directing audience questions to the most appropriate speaker, for instance.

During rehearsals, be supportive, but constructively critical, of each other's content and delivery style. Sometime before the dress rehearsal, go out into the room, sit at the back, and see whether you can hear everything each speaker is saying. Pretend you are an audience member who does not already know a lot about the project. Do the slides and the flow of information make sense? Are the points easy to understand? Does each speaker provide transitions between important points? At the dress rehearsal, of course, the entire team should be up front, and you should all be practicing how to "pass the baton" to each other. Remember to remind the audience of the name of the speaker who will be speaking after you; no one is going to remember an entire list of names on a title slide. Enjoy the synergy of presenting with a group. You will find that the audience will also enjoy that energy. Your tone with your teammates should be natural and relatively informal, and small touches of humor may be used here and there. Ask at least one non-team member to watch a rehearsal and comment about the humor to ensure that it is appropriate and relevant.

Tips for Presenting with a Team

Preparation

- Share contact information at the first meeting. Stay in touch!
- Divide up the preparation according to individual strengths:
  - Who is the best slide designer?
  - Who knows the best sources of information for each topic?

- Divide up speaker roles according to individual strengths:
  - Who knows most about each section or topic?
  - Who is the most upbeat?

- Rehearse often, and seek constructive criticism.
- Work on transitioning smoothly between sections and speakers; do not get in each other's way. Does it make sense to have one person running the computer projection?
- Practice answering questions the team has brainstormed.

Delivery

- "Pass the baton" to each other by mentioning the next speaker's name. Consider previewing what he or she will discuss.
- Answer those audience questions with which you feel most comfortable. If no one can answer a particular question, offer to find out the answer and get back to the questioner later.
- Stay upbeat and enjoy the synergy of presenting with colleagues!

SUMMARY

- Planning an effective presentation means thinking about your audience and their level of familiarity with and interest in your subject matter. You should design visuals and adjust your language to make technical concepts and data clear to that audience.
- Testing your presentation on a sample audience is a great way to prepare. Ask for feedback from the test audience-members on your voice, delivery, and visuals. Was your talk interesting to them, and did they understand everything?
- Visuals for a talk should generally be bigger, with less text than you think, so that people can read them easily from the back of the room.
- Try putting your main point as the title of each slide. Use a sentence and provide as much visual evidence as you can.
- You can use nervousness as a source of energy to make your presentation more vivid and keep yourself alert and on the ball.
- There are many strategies for controlling the level of nervousness that threatens to overwhelm you. Learn which strategies work best for you, and practice using them before and during your delivery.
- Preparing for a team presentation requires rehearsal with your teammates, not just by yourself.

PROBLEMS

7.1. Along with several other students (if possible), attend a talk given at your college or in the community. Take notes on the speaker's delivery, the visuals shown, and your general reaction to the tone and content of the talk. Answer the following questions:

> Was the talk interesting? Did it hold your attention?
>
> Did you understand all the information?
>
> Did the visuals enhance your understanding? If so, how?
>
> How did the speaker's voice and body language enhance or detract from your understanding or your interest?

7.2. If you attended the talk in Problem 7.1 with at least one other student, compare your notes with the other students'. Answer the following questions:

> Did you all agree on which aspects (visuals, language, body language, voice, etc.) were most effective?
>
> What differences do you note in your reactions to the talk? Can you postulate any reasons for those differences?

7.3. Make an outline for a presentation you will deliver in one of your classes. Begin the outline by summarizing in 30 words (or less) your objective in giving the talk. Phrase your objective using an active verb, e.g.: "In this talk, I want to *demonstrate/persuade/inform/show* my audience *something*." Show the outline to your instructor or another student for feedback.

7.4. Pick your favorite design template in MS Powerpoint (or another presentation software package) and create a slide (not the title slide). Try to use a projector to look at the slide on a big screen (rather than on your computer screen). Ask a friend to look with you. Then, answer these questions:

> If your slide has text on it, does the design get in the way of reading the text?
>
> If your slide has a visual, does the design get in the way of seeing and understanding the visual?
>
> If you use this design, which slide formats or layouts will you need to avoid?

7.5. Using MS Powerpoint or another presentation software, create a slide with nothing but red, 24-point text. Then project the slide on a big screen in as big a room as you can find. How close to the screen do you have to sit to be able to read the red text?

7.6. Along with a friend, find a big room, or preferably, an auditorium somewhere on campus or in your town. Ask your friend to sit in the back of the room while you stand in the front or on the stage. Talk in a normal voice. If your friend cannot hear you very well, ask him or her to keep moving closer to the front. How close to the front does your friend have to move to be able to hear you easily?

7.7. True or false? I should work hard to eliminate all traces of nervousness in myself before I give a talk.

7.8. The most important quality in giving a successful presentation is which *one* of the following?

> Showing only one visual per minute (on average)
>
> Never saying "uhm"
>
> Being enthusiastic
>
> Always remembering to step to the side of the big screen

FURTHER READING

Alley, M.; Schreiber, M.; Ramsdell, K.; Muffo, J. "How the Design of Headlines in Presentation Slides Affects Audience Retention." *Technical Communication*, Volume 53, Number 2, May 2006 , pp. 225–234.

Gurak, L.J. 2000. *Oral Presentations for Technical Communication*. Allyn & Bacon: Boston. This is probably the best book available that is devoted entirely to technical presentations; most publications on presentations are overgeneralized and geared to the sales pitch or after-dinner situation. Presentation of complex technical information requires much more thought about the background of your audience and the best way to display data. Chapters 16, 17, and 18, on visuals in presentations are particularly helpful.

Houp, K.W., Pearsall, T.E., Tebeaux, E., and Dragga, S. 2002. *Reporting Technical Information*, 10th ed. Oxford University Press: New York. This venerable textbook, now in its 10th edition, covers all aspects of writing and presenting technical information. The chapters on techniques and applications of writing are excellent, and of all the many technical-communication textbooks out there, this one does the best job on presentations. Make sure to read Chapter 19, "Oral Reports."

Tufte, E.R. 2003. *The Cognitive Style of PowerPoint*. Graphics Press: Cheshire, CT. This pamphlet argues that overuse of PowerPoint has led to a sort of dumbing down of business and professional audiences. Relying on bullet points to convey complex messages, Tufte argues, allows the speaker to avoid his or her responsibility to analyze and describe complexity—the "connective tissue" between those bullets that is the real information.

CHAPTER

8

Producing Engineering Documents: The Final Product

8.1 INTRODUCTION

As an engineer, you will produce many different kinds of documents, including working drawings, specifications, and technical instructions. But the majority of documents you will produce in your career falls into the category of reporting on results or on plans for obtaining results. These sorts of documents generally use sentence-and-paragraph format to convey information to a variety of readers including: your boss, your boss's boss, upper management, clients, subcontractors, suppliers and vendors, public agencies, and sometimes the press. Understanding the expectations of those readers about format as well as content will be critical to success. This chapter presents samples of these most common types of engineering documents, with annotations that explain why each is a good example of its type. The focus here is on the big picture: the way the document is designed and put together to convey the information.

This chapter also describes the document design and production process, for which you may eventually have abundant help in the form of technical writers and editors, but which may at times need your guiding hand. Every technical professional should know how to set up documents for publishing in various forms (hard copy, soft copy, typical digital forms, etc.) and should have some idea of the time and resources required to publish and then disseminate documents.

8.2 DESIGNING PROFESSIONAL DOCUMENTS

Once you have done the research, gathered and organized the information, written (probably) several drafts and revised several times, you are ready to produce the document in hard copy or digital form (often both). This section offers general guidelines for designing readable pages. Most of these guidelines apply to both hard and soft copy, although the end of this section outlines some of the design differences to keep in mind.

You do not need to think of yourself as a graphic designer to realize that two elements are common to all document pages: text and white space. The

Objectives

By reading this chapter, you will learn the following:

- principles for designing professional-looking documents;
- standard formats for the most common engineering documents;
- how to write correspondence;
- how to write a research proposal;
- how to write a progress report;
- how to write a feasibility report.

You will also be able to study annotated samples of the following types of engineering documents:

- memos;
- research proposals;
- progress reports;
- feasibility reports.

way that you arrange those two elements makes some pages more attractive and easy to read than others. White space is the most underappreciated design element. And yet, using white space is the easiest way to set off important words, phrases, graphics, and beginnings of sections—any textual or graphical element of importance.

Use white space to set off important or "different" items:

- figures and tables (including titles);
- titles of sections;
- headings and (some) subheadings.

Using text as a design element means considering and making decisions about justification, indentation, line spacing, and the size and type of font. Those are some of the decisions that affect the look and readability of your text. When choosing a type of font, remember that the commonly used fonts break down into serif and sans-serif fonts. Serifs are the little "feet" appended to the edges of some letters. Times is an example of a serif font. Sans-serif fonts do not have these little marks; the letters are more blocklike. Arial and Helvetica are common sans-serif fonts. Figures 8.1 and 8.2 display the same two paragraphs with two different sets of guidelines for spacing and fonts. In Figure 8.1, the text is fully justified, a line is skipped between paragraphs, and a sans-serif font is used (Arial). Both samples have the same size font (10) and use $1\frac{1}{2}$-line spacing.

In Figure 8.2, the text is left justified (with a ragged right margin), paragraphs are indented (so no extra line is skipped between them), and a serif font is used (Times New Roman).

Which scheme is better? The answer usually depends on what the reader needs to do with the document. For text that needs to be projected (as in a presentation) or read from some distance or under unusual circumstances (such as out in the field), a sans-serif font works best. For text that is long (more than a page) and meant to be read up close, some research suggests that a serif font is best, because readers can distinguish more quickly between the letters. Similarly, some research shows that fully justified text, while it looks very professional, is not as easy to read in longer documents. Readers who turn away from a page can find their place again more easily, it is thought, with a ragged right margin. For two-column formats, however, full justification makes for a much cleaner look.

The scope of my project will involve comparing different wall types, including single-, staggered-, and double-stud walls, constructed with a combination of building and sound-control materials. These materials will include gypsum board, resilient channels, wall cavity sprays, thin polymers, insulation, and sound-deadening boards. The different wall types are shown in Figure 1.

I will not consider noise exterior to the apartment. I will assume that the apartment complexes are two to three stories and of a generic design, as given in Figure 2. In addition, I will only be analyzing the section of wall given in Figure 2 to evaluate the different wall types (called the "party wall," through which most interior noise passes).

Figure 8.1
Paragraph using full justification and Arial 10 font.

The scope of my project will involve comparing different wall types, including single-, staggered-, and double-stud walls, constructed with a combination of building and sound-control materials. These materials will include gypsum board, resilient channels, wall cavity sprays, thin polymers, insulation, and sound-deadening boards. The different wall types are given in Figure 1.

I will not consider noise exterior to the apartment. I will assume that the apartment complexes are two to three stories and of a generic design, as given in Figure 2. In addition, I will only be analyzing the section of wall given in Figure 2 to evaluate the different wall types (called the "party wall," through which most interior noise passes).

Figure 8.2
Paragraph using ragged right margin and Times New Roman 10 font.

Visual Elements of a Page

Text
Headings and titles
Graphics
Simple separators like lines and boxes
White space

Get to know your word processor well so that you can control elements such as spacing, placement of page numbers, and size and placement of headings. If you learn to set up style sheets in your word processor, you can let your computer "remember" what size fonts to use for text, titles, and headings for each document; where to put the page numbers; how much white space to put around headings; etc. In Microsoft Word, the Style command is in the Format menu.

Designing Pages: A Few Tips

- Use the same design for all pages that contain the same kinds of information (e.g., beginnings of sections).
- Use a maximum of two typefaces. Serif fonts such as Times are slightly easier to read in long documents.
- Text with a ragged right margin is slightly easier to read in long documents than is text with a fully justified margin.
- ALL CAPS ARE HARD TO READ; USE THEM SPARINGLY.

Headings play a key role in helping the reader identify information quickly. Use plenty of white space around headings (more space for bigger headings), and check carefully that you have kept levels of headings consistent. After the title of the document and the titles of sections, the major headings are level #1, the sub-headings are level #2, etc. Be sure to make the look of each heading reflect its level. Thus, if "Introduction" and "Conclusion" are both major headings, their size, font, and format should be exactly the same. Figure 8.3 shows some examples of different options for keeping headings consistent.

Headings
Option One: Use Size and Position to Differentiate

Heading Level One   [14-point Times New Roman]

Heading Level Two   [12-point Times New Roman]

Heading Level Three. Might be run in with text. If you number the heading, do not use bullets points.

Be consistent with capitalization.

Run-in headings usually need to be in boldface or italics to set them off from the text.

(a)

Headings
Option Two: Use Numbering to Differentiate

1. Heading One

1.1 Heading Two

1.1.1 Heading Three

This example uses position, as well as numbering, to differentiate the headings.

(b)

Headings
Option Three: Use Font Styles to Differentiate (Example #1)

Heading One   [Times New Roman 14 bold]

Heading Two   [Times New Roman 12 unbolded]

Heading Three. Italics help differentiate heading from body text.

This option also generally uses size as well as style to differentiate between heading levels.

(c)

Headings
Option Three: Use Font Styles to Differentiate (Example #2)

Introduction   [Tahoma 12]

    This is the introduction, etc.   [Times New Roman 10]

You might use one font type for headings and another for text. *Never use more than **two** font types* in a technical document (except for manuals).

(d)

Figure 8.3 (a–d)
Different options for keeping headings consistent in format and level.

Since size is the most intuitive reflection of importance, choose size first as a way to indicate level of heading or importance, and then add on other differentiating characteristics. Do not use bullet points in headings.

To avoid last-minute chaos, make decisions about headings and text design early in the process of writing. Learn to set up style sheets in your word processor so that all headings of a certain level will be consistent. If you are writing as part of a team, all members will need to agree about these stylistic decisions. The style sheet checklist in Figure 8.4 should help you remember all of the style issues that will need

Style Sheet Checklist

† **Document**
  – Paper type and cover page
  – Single sided or double sided?
  – Margin width

† **Sections**
  – Start on right-facing pages?
  – Use line or other graphic device to divide?

† **Headings and Subheadings**
  – Position
  – Font: size and characteristics (e.g., bold, italics?)

† **Main Text**
  – Font (serif or sans serif)
  – Paragraph indentation, or skip a line between left-justified blocks?
  – Line spacing (single, 1½, or double)

† **Page Numbering**
  – Where on page?
  – What font or type?
  – Different for different sections?

† **Tables and Figures**
  – Integrated with text?
  – Where do labels go?

† **References**
  – Citations in parentheses in text?
  – Reference list goes where?

† **Abbreviations, Acronyms, Equations**
  – Include glossary? Where?
  – Acronyms spelled out where?
  – Mathematical variables explained where?

Figure 8.4
Standard checklist for a document's style sheet.

resolution (e.g., Where should page numbers go?). If you resolve these issues sooner rather than later, you save yourself headaches in the production as well as the writing process. If each person is supposed to write 8–10 pages for his or her section, for instance, what happens when one writer is using 8-point Times and therefore ends up producing lots more text than the person using 12-point Times? You have a much more difficult editing and cutting job—that is what happens.

More Tips for Designing Pages

- Find out whether your organization has a style guide. If so, use it!
- The team producing a document needs to create style guide if one does not already exist.
- For journal papers or conference proceedings, check the Information for Contributors or Call for Papers sections (usually in the front of the journal) for style guidelines.

Standard fonts for engineering reports are Times and Times New Roman 11 or 12. Some companies prefer Palatino or other fonts. Always check with your professor, company, or agency to see whether a style guide (even an informal one) exists.

8.2.1 Disseminating Documents Digitally

If you need to disseminate digitally a word-processing document that you do not want altered (such as a final report or perhaps your résumé), it is generally a good idea to convert the document to PDF, or portable document format. Some word-processing programs, such as Microsoft Word, allow you to convert documents to PDF; you can also convert documents using Adobe® Acrobat® software. Your file can then be opened using Adobe® Reader®, which is available free of charge at www. adobe.com. PDF has become the standard format for exchanging documents, since PDF files cannot be altered as easily as files in other formats, such as Microsoft Word.

ASSIGNMENT

Take two pages of a document you have already written, and try setting up styles for it. In Word, define the styles for the different levels of your headings: Choose the type and size of font (bold? italics? underline?).

Choose the spacing before and after the headings. Define the style of your "normal" text. How does your document look when you apply those styles? Now change some of the styles and see which ones you think make the document look better. More importantly, show a couple of different styles to a friend, and ask which one he or she likes better.

FURTHER READING ON DESIGNING DOCUMENTS

Williams, R. 1994. *The Non-Designer's Design Book*. Peachpit Press: Berkeley, CA. This little book is the best guide to document design for folks who are not graphic designers.
Williams, R., and Tollet, J. 1998. *The Non-Designer's Web Book*. Peachpit Press: Berkeley, CA. A follow-up to the previous reference, this volume provides helpful guidance to the novice about how to create and post a website.

8.3 SAMPLE CORRESPONDENCE

This section presents an internal memo from the head of quality control (QC) at a wood-truss manufacturing company to a supervisor who is not an engineer. The company has just received a shipment of wood from the supplier, and uncharacteristically, the wood has a moisture content above the fiber-saturation point. The supervisor (Ellen Austin) majored in business and believes the wet wood would be fine to use for the current construction job. George Lumberman, the QC engineer, must explain the technical consequences of using the wet lumber.

8.3.1 A Word about Using the Sample in This Section

This section presents a model of a good letter for you to use. The model may not, however, fit the purpose or the requirements of a document you are called upon to write. Always ask your boss or instructor for samples to look at. Don't then copy them slavishly by using a fill-in-the-blanks approach, but rather notice the type of information presented in each document or section and decide whether your information fits that type.

LONGHORN LUMBER INTRAOFFICE MEMO

DATE:	11/18/2003
TO:	ELLEN P. AUSTIN, MANAGER
CC:	MARIA G. JONES, PRESIDENT
FROM:	GEORGE G. LUMBERMAN, QUALITY CONTROL
RE:	NUECES CONSTRUCTION PROJECT

It has come to my attention that the Southern pine lumber we recently received from Peachtree Lumber Supply has a moisture content above the fiber-saturation point. I was quite surprised, as Peachtree has an excellent track record of shipping well-seasoned, high-quality lumber. I have contacted Peachtree, and they regret the mistake (citing a recent roof leak in their warehouse) and have offered to send a new shipment free of charge, contingent on the return of this one. This is a very generous offer, but somewhat inadequate. It would take 10 days for the new shipment to arrive, and this lumber was intended for use in the Nueces construction project. As you know, the Nueces project is already behind schedule. The workers will be ready to install the trusses tomorrow, and no further work can be done until they are in place. However, it is highly preferable to halt construction for 10 days and await the new shipment than to use the lumber we currently have on hand.

Using lumber with a moisture content above the fiber-saturation point would be catastrophic for the structural integrity of this building. Wood is made of thin-walled tubes, which are also called cellulose cells or fibers. When wood absorbs water, these fibers become saturated with water, swell, and move apart. The fibers give wood its mechanical strength; as the fibers move further apart, the density of the fibers decreases and the wood loses strength and stiffness.

The relationship between the strength of wood and its moisture content is demonstrated by a test that I performed on the lumber we received this week and some well-seasoned lumber left over from a previous project. Each beam was placed between two supports, and a load was applied at a constant rate to the center. The test results are tabulated in the upcoming table. In both tests, the dry lumber sustained significantly larger loads before failure. The structural design for the Nueces project demands that the trusses be much stronger than the wet beams we tested. Furthermore, the displacements recorded for the wet beams are much larger than for the dry ones. The design cannot accommodate such large displacements.

Sample description		Maximum load (lbs.)	Displacement (in.)	Strength ratio (dry/wet)
Test 1	Dry	2557	1.14	1.707
	Wet	1498	1.60	
Test 2	Dry	1905	1.08	1.922
	Wet	991	1.50	

Therefore, my recommendation is to *halt construction* until the new lumber shipment arrives. An alternative you may have considered is to let the wood we have dry out. However, this process is likely to take longer than 10 days, given the current damp weather conditions. Another alternative is to investigate other suppliers who could fill our need more quickly.

Under no circumstances should we use the wet wood. Not only would using this wood cause an unacceptable safety hazard, it would be catastrophic to our reputation.

NOTES ON SAMPLE MEMO

Notice that in correspondence (especially in internal memos), personal pronouns can be used. Because all correspondence is written from one person to another, these pronouns are acceptable, except in the most formal, legalistic letters. Notice also how the author, who is an engineer, gently educates the recipient, who is *not* an engineer, about the structural integrity of wet vs. dry wood. It is not unusual for middle managers in engineering companies to have business backgrounds rather than engineering degrees. Even with engineering degrees, some supervisors have been away from hands-on engineering long enough that they do not remember or have not kept up with technical work in their field. Understanding your audience is an important part of communicating technical information effectively. Even when the audience is your boss, do not assume that person has the same background that you do, and think about what technical explanations may help him or her make good business *and* engineering decisions.

In this memo, the most important points are made at the beginnings and endings of paragraphs. The fact that the wood is wet is given in the very first sentence. The first paragraph then ends with the information that halting construction is preferable to using the wet lumber for the current project, and the next paragraph begins with a firm statement about the unacceptable risk to structural integrity of using the wet wood. The word "catastrophic" should not be used very often, but when you need to use it, do not bury it in the middle of a paragraph. The final paragraph begins with the recommendation not to use the wet wood, and the memo ends with a warning of the consequences of not following that recommendation. The final paragraph has only two sentences—an uncommon occurrence in technical writing, but justified by the critical nature of the information imparted.

Figures and tables are handled differently in correspondence than in other technical documents. Notice that the table has no title or number. Generally, the few graphics in letters or memos are explained so thoroughly that they do not need formal labeling, except if they are included as attachments and therefore might be photocopied or read separately.

Generally, memos and letters are single spaced.

8.4 SAMPLE RESEARCH PROPOSALS

This section presents two sample research proposals. The first proposes a consulting project to evaluate different materials for soundproofing an apartment building. The comments in boxes highlight the function of each section within this proposal. The second proposal is a response to a Request for Proposals (RFP) circulated by the New Product Development Group of an electronics company. This second proposal offers a different organizational scheme and includes a Literature Review. When responding to an RFP, you must use the organization and format requested or your proposal probably won't even get read. See Section 8.6, Sample Feasibility Report for more information on structuring longer documents.

8.4.1 A Word about Using the Samples in This Section

This section presents two models of good research proposals for you to use. Neither of the models may however, fit the purpose or the requirements of a document you are called upon to write. Always ask your boss or instructor for samples to look at. Don't then copy them slavishly by using a fill-in-the-blanks approach, but rather notice the type of information presented in each document or section and decide whether your information fits that type.

Sample Proposal 1

Relyon Engineering, Inc.

Proposal to Determine Improved Methods of Sound Insulation

Summary

The residents of Gaylord Apartments in central Austin have complained to the owner, Louise Smith, about the lack of proper soundproofing in the building. Ms. Smith is aware that several tenants intend to relocate because of high noise levels, and she is concerned about the possibility of also losing new business if she does not address the problem.

Ms. Smith has requested the services of Relyon Engineering, Inc., to propose and evaluate alternatives for reducing noise between apartments. These alternatives include improvements to insulating materials, alterations to flooring materials, or some combination of the two. We will evaluate potential solutions based on their effectiveness in reducing sound transmission, as well as their cost per apartment, construction time, and aesthetics. Upon completing our research, the Relyon team will present the findings to Ms. Smith and recommend the most appropriate solution to the noise problem.

The estimated time frame for research and evaluation of the sound problem is 11 weeks. We will deliver a progress report to the client on March 24, 2003, and a final report containing the findings and the recommended solution by May 4, 2003. The estimated cost of Relyon Engineering's services for this evaluation is $1,400.00.

Definition of the Problem

In recent months, a large number of residents have complained about the level of noise coming from other apartments in the building. Although noise control is adequate between adjacent apartments on the same floor, excessive sound is transmitted through ceilings and floors and creates a major disturbance for the tenants. In the past year, approximately 20% of the building's residents did not renew their leases, and 50% of those tenants cited high noise levels as a factor in their decision.

Potential causes of this problem include inappropriate choice of flooring materials, a lack of proper insulating materials in apartment ceilings, and structural damage to the ceilings. Hardwood floors, which are in all of the building's apartments, transmit more sound than carpeted floors. Various types of

Clear and specific statement of the problem

Statement of purpose

Possible solutions have been narrowed down

Criteria for evaluating solutions described

Answers to managers' questions: How long will project take? How much will it cost?

Expansion of statement of the problem

insulating materials can be placed inside or on the exterior of ceilings to help absorb noise; the absence or underutilization of such materials could be a factor in the excessive noise transmission. Structural damage such as small cracks can also significantly affect the ceiling's ability to act as a sound barrier; in an older building such as Gaylord Apartments, such damage would not be unexpected.

In order to improve the quality of life for the remaining residents, as well as to increase the market value of these apartments, Ms. Smith feels it is necessary to implement a solution before July, when most leases are up for renewal.

Scope of Project

Our team will consider two basic methods to increase the sound insulation in the apartment: carpeting the floor in the apartments and adding insulating materials to the ceilings. Possible solutions could also include a combination of these techniques. Although we have also discussed structural damage as a potential cause of the problem, any repairs of this nature would be too time consuming and disruptive to residents and therefore are beyond the scope of this investigation.

Expanded description of possible solutions

The general scope of the project will be to more fully assess the problem, determine which specific materials (flooring and/or insulation) we will consider using to solve the problem, and analyze each possibility against our criteria, which are described in the next section. Our research will include sound measurements inside the building to help us quantify the current noise levels and resident surveys to gain more information about what improvements they want or expect to see. Consultation with an acoustical consulting firm and companies that produce insulating materials will provide a basis for more specific alternative solutions. We will then gather additional information from textbooks, journal articles, and product specifications to help us determine which solution to recommend to the client.

Methodology used to study problem and evaluate solutions

Our final report will recommend the strongest solution of those studied. We will present these solutions to the client in both oral and written form. The oral presentation is scheduled for April 23, 2003, and we will submit the final report on May 4, 2003.

Reiteration of project's time frame

Solution Criteria

The alternative solutions will be evaluated according to the criteria outlined next. These criteria are listed in order of priority, from highest to lowest, as designated by the client.

Whole section is a detailed description of criteria to be used to evaluate possible solutions

Effectiveness in Reducing Noise Transmission

The client has asked that we determine an appropriate threshold for noise transmission, based on the preferences of residents. We will measure each solution in terms of noise reduction (NR), a quantifiable term expressed as a function of the transmission loss (TL) of a barrier (defined as the ratio of the sound reradiated by the barrier to the sound absorbed by it, in decibels), the area of the barrier, and the ability of the receiving room to absorb sound.

Cost of Implementation

The owner would like to spend no more than $1,500 per apartment to implement the solution. This cost includes materials and labor, but is separate from the consulting fees for Reylon Engineering's recommendation report.

Time of Construction

Most new resident apartment leases begin July 1, and the client has specified a construction completion date of June 15 in order to prepare the apartments for their new tenants. Since vacating tenants must leave by June 1, renovations must be completed in the two-week period from June 1 to June 15.

Aesthetics

We will consult residents as to their aesthetic preferences among the various solutions involving carpeting and ceiling insulation. We will also survey residents to determine how important aesthetics are relative to effectiveness in reducing noise, using a simple numerical scale to rank their preferences. These rankings will help us determine how much consideration to give aesthetics in our final recommendation.

Proposed Procedure

We will begin our research by investigating the apartment building itself. We will obtain the original building plans and determine what materials are presently acting as sound barriers in the ceilings, if any.

Step-by-step outline of tasks necessary to accomplish goal of project: recommend best solution to the noise problem

We will also visit the apartments and determine the amount of sound being transmitted between apartments through the ceilings and floors, using standard equipment for measuring decibel levels. This process will help us determine the current noise reduction (NR) between apartments and provide a baseline with which we can compare the effectiveness of potential alternatives. Surveys of the residents will provide us with information about what constitutes an acceptable level of noise, as well as what aesthetic preferences they have, if the ultimate solution involves altering the flooring material. Based on these data, we will set specific targets for the NR and aesthetics criteria.

Once we have clarified these goals, we will consult professionals currently working in the area of residential sound insulation, including the acoustical consulting company Nelson Acoustical Engineering, Inc., and Victor Gomez from Flour Daniel, Inc., to help us develop a more specific list of possible solutions. We will also contact the insulating-materials companies Netwall Noise Control and Building Green Products to gain information about particular products that could aid in improved insulation and noise reduction. In all of these cases, correspondence will be mainly via telephone and in-person office visits.

After finalizing our list of possible solutions, we will consult the appropriate technical journals, textbooks, and product specifications to estimate each alternative's effectiveness in reducing transmitted noise (NR), as well as the material and labor costs involved. To evaluate the amount of time necessary to implement each solution, we will consult area contractors who have experience with the specific renovations we are considering. Finally, survey data gathered from residents at the beginning of our project will allow us to factor the aesthetic criterion into our recommendation.

All feasible solutions will be presented directly to the client in an oral presentation, followed by a written report of the results of the study, which will include material layouts and a cost breakdown, as well as cost estimates for the entire project.

Schedule

The proposed work schedule for the project is as follows:

> **A page break is inserted here so as to fit Table 1 on one page. Never break a table over two pages!**

Table 1. Project Schedule

Activity	2/22	2/26–3/05	3/05–3/12	3/12–3/19	3/19–3/26	3/26–4/09	4/12–4/24	4/23–5/04
✓ Analyze building plans	←→							
✓ Perform sound measurements		←→						
✓ Survey residents		←→						
✓ Consult sound engineers			←→					
✓ Consult materials companies				←→				
✓ Evaluate solutions					←→			
✓ Deliver progress report					3/26			
✓ Prepare report draft						←→		
✓ Submit report draft to supervisor						4/09		
✓ Review draft and prepare oral report							←→	
✓ Deliver oral report to Harriet Romo							4/24	
✓ Prepare final report								←→
✓ Submit final report								5/04

> **Table title in correct location**

Budget

The estimated total cost of our team's services is $1,400.00. This total includes all research, funding for collaboration with other firms, and preparation of reports, as shown in the following cost breakdown:

> **Narrative justification of budget**

Table 2. Project Budget

Activity	Number of Hours	Hourly Rate	Cost
Research and analysis	15	$35	$525
Correspondence with other firms	5	$35	$175
Cost estimating and comparisons	5	$35	$175
Presentations and reports	15	$35	$525
		Total Cost	**$1,400**

> **Table gives quick visual overview of budget**

Qualifications

The project manager has a good working knowledge of the fundamentals of architectural, structural, and mechanical engineering. She also has experience working in a large civil-engineering firm.

Jane Daniels

Education: *Bachelor of science in civil engineering*, the University of Texas at Austin, to be awarded December 2005.

Experience: *Intern*, Turner Collie and Braden, Inc., June 2003 to August 2003.

> Qualifications of researcher doing the work

Sources of Information

- Stein, Benjamin, and John S. Reynolds. 2000. *Mechanical and Electrical Equipment for Buildings*. 9th ed. New York: John Wiley & Sons, Inc.
- *Journal of Sound and Vibration*, London.
- Victor Gomez, P.E., Flour Daniel, Inc., Austin, TX.
- Gina Davis (interview, March 23, 2002), Nelson Acoustical Engineering, Inc., Austin, TX.
- Netwall Noise Control, Austin TX.
- Building Green, http://www.buildinggreen.com/index.html (accessed March 2002).
- Louise Smith, owner, Gaylord Apartments.

> This reference list is not formatted according to any accepted conventions; it is simply an indication of where this student will look for information. Ask your instructor whether you should use a formal references list.

Sample Proposal 2

Date:	March 31, 2008
To:	Dr. Fagelson, Vice-President, New Product Development
From:	GPU Research team
Subject:	A proposal to research the effectiveness of using graphics processing unit (GPU) in supercomputing applications

1.0 INTRODUCTION

The following document is in response to the internal Request for Proposals put out by the New Product Development group on March 22, 2005. The purpose of this proposal is to obtain approval to research the use of graphical processing units (GPUs) as co-processors for a more cost-efficient method to increase and optimize computational power.

Today's supercomputers require huge amounts of money, power, and space to maintain, something we are unable to afford due to the recent economic downturn. Because our company utilizes a massive amount of servers to support employees and customers alike, optimizing computational performance with the use of GPUs may save company resources. Though GPUs were originally designed to process large amounts of graphics' data, their capabilities may also be used to perform and accelerate non-graphics computations, leaving the CPU free for other tasks [1]. However, to effectively evaluate the feasibility of using GPUs, further research must be conducted into the efficiency of the processor in a variety of applications and hardware structures.

The proposal includes a brief project description, a preliminary research report, the approach the group is taking on the project, a timeline of the project, and an outline of the final report. A thorough evaluation on the feasibility of the use of GPUs will be presented in the final report, pending further research.

2.0 PROJECT DESCRIPTION

This project will evaluate the effectiveness of using a GPU as a co-processor to the CPU when performing computations with large date sets. This section presents the problem that drives this project's investigation, a description of the proposed solution, a description of the project activities, and an overview of the material that will be included in the final report.

2.1 Project Problem Statement

Research facilities around the world need a more efficient method to compute massive amounts of calculations. Current methods of computing are too slow and costly to maintain.

2.2 Description of Research Topic

To solve the need for better computational efficiency, we propose use of a GPU co-processing system as opposed to other alternatives such as using many CPUs in parallel or using advanced cooling systems. Implementing enough CPUs in parallel to equal the computing power of a single GPU comes at a much higher power and size cost. Advanced cooling systems also increase performance of existing systems but add cost and size to the overall system. The GPU co-processor solution uses hardware that most computer systems already have; therefore, this solution does not significantly increase cost, size, or power consumed. Both alternative methods above improve speed, but when considering computational efficiency, the GPU solution appears to have the most potential gain with the least amount of cost.

2.3 Description of Research Activities

We will search for experiments with existing GPU systems and evaluate each system's ability to improve computational efficiency in a variety of applications. Our main sources of relevant information will be journals, such as the *Proceeding of IEEE*, and results of studies published by various universities and research labs on the Web. Our goal is to determine the positive and negative attributes of a GPU system and objectively compare the results to current technology standards. Some key difference we are looking for when processing large data sets are changes in speed, cost, and size. Naturally, using the GPU as a co-processor has both hardware and software limitations, and those limitations will be fully evaluated for their effect on the project's overall feasibility. The performance attributes of the GPU system will be investigated, as well as possible applications for this technology. For the existing successful applications, we will determine what strategies, such as hardware implementation or programming style, impacted the success of the outcomes. We will also investigate the market for this technology. Through research and evaluation, we will determine if using a GPU as a co-processor increases performance at reasonable costs and where this technology can be applied successfully.

2.4 Purpose of the Final Report

The purpose of the final report is to inform management of our research and conclusions as to the GPU's effectiveness in supercomputing, feasibility of implementation, and market need. The feasibility of the system will be conveyed in different sections of the final report, which include the benefits and limitations of a GPU co-processing system, different applications this system can be used in, and what strategies are used in successful applications of this system to maximize performance (refer to Appendix A for a complete outline of the final report). Additionally, the final report should inform management of the potential gain in computational efficiency while outlining the potential risks involved with investing internal resources towards further research and development. In addition to providing a recommendation to the New

Product Development Group, the final report will present a detailed evaluation of the GPU co-processing system's ability to meet the needs for a more efficient computational solution.

3.0 LITERATURE REVIEW

The team found several articles in preliminary research on the computing power of the GPU in both graphical and non-graphical computations. These articles are summarized below

3.1 Reference Articles

"GPU Cluster for High Performance Computing" [1]

This paper compares the speed of a cluster of 32 systems that contain GPUs and CPUs, and of an equivalent system using just CPUs in specific applications. The price/performance ratio and evolution rate of the GPU are discussed. The results of these findings show that adding GPUs significantly speeds up the cluster by at least four times the equivalent system without GPUs for this application. These findings present a unique implementation of the GPU in a cluster, a possibility that needs to be further researched.

"NVIDIA GeForce 256 Review" [2]

This online article gives a review of the GeForce 256, one of the first graphics cards to use a GPU. Containing over 22 million transistors, the GeForce 256's GPU is capable of processing billions of calculations per second. Such capability frees up the CPU to enhance the physics of game animation and the logic in advance artificial intelligence. This article is another source citing the tremendous capabilities of the GPU, even in its early stages of development.

"Scout: A Hardware-Accelerated System for Quantitatively Driven Visualization and Analysis" [3]

This article outlines the creation and applications of GPU co-processing system called Scout. These applications include multi-dimensional transfer functions, multivariate visualizations, and deriving multi-variable visualization fields. Eliminating the need to transfer calculated data from the CPU to the GPU is one benefit of performing calculations with the GPU instead of the CPU in visualization problems. Another improvement of this GPU co-processor system is the decrease in calculation times. However, the calculated times do not include the initial data-loading stages, which can take up to one second for a large data set. In addition to the lengthy setup time, the GPU hardware supports a limited instruction set (used to write programs) and limits the number of temporary variables used in calculations. If the GPU can be leveraged to perform advanced analyses, we could see a dramatic decrease in computation times.

"Computer Vision Signal Processing on Graphics Processing Units" [4]

This article examines the performance of the GPU co-processor for computer vision. Computer vision is a form of image analysis that takes a two-dimensional image and converts it into a mathematical construct. Using fragment shader programs, the GPU performs up to 3.5 times faster than the regular CPU and shows no apparent hardware bottlenecks that may hinder the computations. This article also cites a multiple GPU design that not only increases the overall memory bandwidth of the system but also prevents the GPUs from contending with each other for access to a shared memory area. Though the study conducted in this article cites several advantages of using CPU for non-graphical computations, it was done only on particular operations rather than full computer vision algorithms.

3.2 Problems In Getting Additional Material

Because this is a relatively new implementation of the GPU, previous studies are not uniform in testing methods, applications performed, and hardware structures of the GPUs tested. Because of this lack of uniformity, these studies fail to compare their results with those of other tests on different computations and hardware setups. However, we can draw our own conclusions from the various testing schemes.

3.3 Additional Sources to Consult

We plan to consult additional journal articles and studies on the performance of the graphics processing unit, GPU, in a variety of hardware structures and computational applications. Also, to conduct market research, online web servers will be consulted for traffic volume and cost.

4.0 PROBLEM ANALYSIS AND APPROACH

This section describes how the research questions will be broken down for detailed investigation and how the research will be conducted and reported.

4.1 Research Techniques

Coordinating the research will require two phases. Phase one will look into what kinds of applications should be run with GPUs to get a fair representation of performance and will develop criteria by which to evaluate the resulting performance. Phase two will investigate the feasibility of these enhanced applications based on their marketability, their problem-solving capability for the company, and their cost/benefit ratio. These tasks involve running tests of the applications using GPU's, evaluating their performance, calculating costs and benefits of the applications, and reaching conclusions about the feasibility for the company of using GPUs.

4.2 Research Subquestions

Some questions must be answered in the final report related to the background, feasibility, and business and marketing aspects of investing in GPUs.

4.2.1 Background

• What is supercomputing?

• What are the problems with supercomputing?

• What types of problems are solved with supercomputing?

• What is a GPU and its original purpose?

• Why use GPUs as co-processors?

4.2.2 Feasibility

• What current applications can be run on a GPU system?

• What kinds of problems can be solved with GPUs?

• What kinds of problems cannot be solved?

• What does the future hold for GPUs?

4.2.3 Business

• How will this technology help the company?

• Can this be marketed to others?

• Are there similar solutions already?

• Will the performance gain outweigh the cost?

• Are the costs of developing software on the GPU worth the performance?

4.3 Presentation of Report

The findings will be presented in a final report as well as an oral presentation. Figures and tables will demonstrate ideas and trends that are vital to the understanding of this technology. A progress report detailing our current status will be delivered midway through the project schedule. See Appendix A for the final report outline.

4.4 Project Collaboration

In order to complete the project, engineers David Ghosti, an expert on GPUs; John Riley, computing research specialist; Brian Aherne, a resource manager; and Jessica Lindford, a market analyst, have been assigned to aspects of the project appropriate to their expertise. David will research background information on GPUs and help edit the final report. Jessica will report on market trends and demands for high performance computing. Brian will research performance tests conducted on the GPU systems, and John will be the guiding editor who compiles the final report. We plan to delegate responsibilities effectively and communicate regularly through email and face-to-face meetings.

5.0 PROJECT SCHEDULE

The project schedule has been arranged so that each week the project team will research one aspect of the GPU system. This arrangement will allow all engineers on the project to understand one piece of it at a time. See Appendix B for the project schedule. Research will begin on April 1, and the final report will be delivered on May 5, 2005.

6.0 CONCLUSION

This proposal seeks permission to research the GPU system's ability to solve the need for more efficient computational power. Our project focuses on investigating the benefits and limitations of the system and evaluating the market need and value of the GPU technology. We believe we can accomplish all of the tasks proposed because we are all diligent workers with a realistic project schedule. The schedule is not only an organizational tool but will hold us accountable to upper management for prompt project completion. This project is a worthwhile investment because it has a large potential gain at a minimal cost.

7.0 REFERENCES

[1] Zhe Fan et. al., "GPU Cluster for High Performance Computing," *ACM / IEEE Supercomputing Conference 2004,* Pittsburgh, PA, November 2004.

[2] Chambers, Mike, "NVIDIA GeForce 256 Review," http://www.nvnews.net/reviews/geforce_256/preface.shtml (21 Oct. 1999).

[3] Patrick McCormick et. al., "Scout: A Hardware-Accelerated System for Quantitatively Driven Visualization and Analysis," *IEEE Visualization 2004 (VIS'04)*, Austin, Texas, Oct. 2004, pp. 171-178.

[4] James Fung and Steve Mann, "Computer Vision Signal Processing on Graphics Processing Units", *Proceedings of the IEEE International Conference on Acoustics, Speech, and Signal Processing (ICASSP 2004),* Montreal, Quebec, Canada, May 17-21, 2004.

Appendix A - Final Report Outline

1. Introduction

   a. Scope

   b. Purpose

   c. Background

2. Technical Background

   a. Super Computing Background

   b. The Original Purpose of the GPU

   c. Trends

3. Other Solutions

   a. Parallel CPUs

   b. Cluster Systems

   c. Advanced Cooling Systems

4. Comparisons to current technology

   a. Requirements

   b. Benchmark-Speed Tests

   c. Power

   d. Costs

   e. Size

5. Benefits And Limitations

   a. Benefits

   b. Limitations

6. Applications

   a. Scientific

   b. Business

7. Market Demand

   a. Governments

   b. Universities

   c. IT Companies

8. Evaluation

9. Recommendation

10. Conclusion

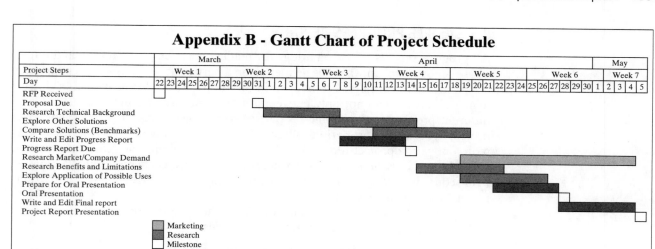

8.5 SAMPLE PROGRESS REPORT

All progress reports provide a bridge for readers between past, present, and future. These reports explain what has happened on a particular project, what is happening now, and what is projected to happen in the future. They also provide a bridge for management readers not only between different phases of a project, but between different projects, each of which is part of some bigger plan or process. For some projects, you will report periodically, as often as once per week; for others, you may write only a midproject report. In all cases, you are helping the reader look ahead toward the successful (you hope) conclusion of the project, whether it is a building being constructed, a manufacturing process being implemented, or a series of experiments testing some new material.

There are many good reasons for taking the time to stop what you are doing and write a project report. Here are a few reasons:

- to reassure clients that you are making progress
- to provide clients with a brief look at some findings
- to give clients a chance to evaluate your work and request changes
- to give you a chance to discuss problems with the project
- to force you to adjust your work schedule to conform to reality

Very often, progress reports are compared with a project's original proposal to see whether the project is keeping to budget and schedule. And, very often, engineering projects deviate somewhat from what is expected. So, a major function of this type of report can be to explain and justify necessary changes and to show how problems (many of them unforeseeable) are being handled. Writing this type of report and presenting it to the client also gives you the chance to solicit help in dealing with problems or with changes requested by the client. The sample progress report shown here has no such problems, but many projects need more money or resources than originally thought.

Here is a sample report on the status of a project to examine various orbital configurations of satellites around the moon in order to help NASA establish continuous communication with the far side of the moon. See Section 8.6, Sample Feasibility Report, for more information on structuring longer documents.

8.5.1 A Word about Using the Sample in This Section

This section presents a model of a good progress report for you to use. The model may not, however, fit the purpose or the requirements of a document you are called upon to write. Always ask your boss or instructor for samples to look at. Don't then copy them slavishly by using a fill-in-the-blanks approach, but rather notice the type of information presented in each document or section and decide whether your information fits that type.

Wentogle Incorporated
Sending You to Outer Space

Progress Report on the Evaluation of Orbit Configurations That Provide Continuous Communication with the Moon for NASA

Project Description

NASA has requested the development of a system to establish continuous communication with the far side of the moon for future scientific missions. Currently, Earth-based communication with the back side of the lunar surface is impossible because of transmission blockage caused by the mass of the moon. On September 13, 2000, NASA requested that Wentogle Incorporated perform an evaluation of possible satellite orbital configurations capable of providing the communication link.

Wentogle's analysis focuses on the optimization of satellite orbital configurations rather than the type of communication satellites to be used for creating the communication network. Three alternative solutions are currently under investigation: two configurations of satellites in circular orbit around the moon and one configuration of satellites in a lunar halo orbit around the gravitational equilibrium point L1.

Complete communication coverage of the back side of the moon is the primary criterion for selecting a configuration. Other selection criteria include the satellite's orbital stability and lifetime, as well as an overall configuration cost estimate, which will be based on the number of satellites required to provide continuous communication.

The project was initiated on October 1, 2000, by John Carson and Andrew Giacobe of Wentogle, Inc., with a budget of $2,000.

Work Accomplished

The following work on this project has been completed thus far.

Satellites in Circular Orbit around the Moon

The geometrical analysis of the circular orbits was performed on systems of satellites in two perpendicular orbit planes. Two separate cases were considered: three satellites per orbit and six satellites per orbit. The configuration for these orbits, shown in Figure 1 and Figure 2, displays the two perpendicular orbits with three satellites per orbit and six satellites per orbit, respectively. Lunar communication coverage provided by the circular orbiting satellites was determined using a geometric analysis based on the altitude of the orbit. For the analysis, the satellites were assumed to be equidistant from each other, thus forming equilateral triangles.

> Figures are referenced and explained in the text.

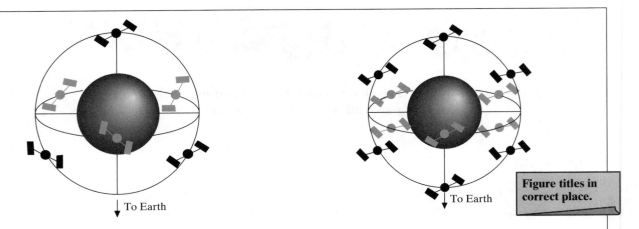

Figure 1. Two groups of three satellites in perpendicular orbits

Figure 2. Two groups of six satellites in perpendicular orbits

Wentogle discovered that the three-satellite configuration provided complete communication coverage when the orbital altitude was a minimum of 3,480 kilometers, which is twice the radius of the moon. The high altitude of the three-satellite system may result in orbital instabilities, so Wentogle decided to research a configuration using more satellites at a lower orbit. The six-satellite configuration required half of the altitude needed for the three-satellite system. The six-satellite altitude is consequently equal to the lunar radius of 1,740 kilometers. The maximum altitude of both orbits is equal to the SOI (Sphere of Influence), which is located at an altitude of 64,380 kilometers above the moon by definition. Above the SOI, Earth's gravitational attraction on the satellites is stronger than the moon's, and the orbits would not remain around the moon.

Satellites in Halo Orbits around the Moon

The communication coverage provided by satellites in orbit around the equilibrium point L1 was determined using a geometrical analysis. The halo orbit of interest is perpendicular to the line between Earth and the moon, as shown in Figure 3. From the analysis, Wentogle discovered that three satellites forming an equilateral triangle in the halo orbit provide communication coverage when the orbital radius is a minimum of twice the lunar radius, or 3,480 kilometers, as shown in Figure 4.

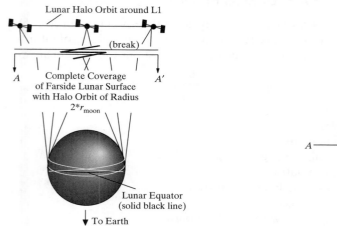

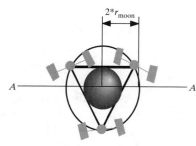

Figure 3. Lunar Halo Orbit with three satellites provides complete communication coverage

Figure 4. Lunar Halo Orbit with three satellites (view of far side of moon, looking toward Earth)

The geometrical analysis was based on a halo orbit with an altitude of 61,500 kilometers above the center of the lunar equator. This altitude corresponds to the distance of the gravitational equilibrium point L1 from the center of the moon. The maximum halo-orbit radius depends on the stability of the orbit and will not be computed until the stability analysis has been completed.

Work Remaining

Based on the geometrical analysis, both the three-satellite and six-satellite configuration of the circular orbit provide complete communication coverage of the lunar surface. However, accurate orbit simulations will be required to determine whether the selected orbit altitudes are stable. Using the models currently under development and performing a geometric analysis, Wentogle will determine the orbital stability and lifetime of the three-satellite and six-satellite configurations. Similarly, once the halo-orbit simulations are complete, their stability will also be analyzed using the computer models. Then, the stability and lifetime data will be compared with the results from the lunar-circular-orbit analysis.

Finally, we will determine the overall maintenance and implementation costs of the configurations. The cost will be determined for the three-satellite and six-satellite configuration, and a comparison will be made between these two options and the option of satellites in halo orbit around the L1 gravitational-equilibrium point. This information will allow us to select the most suitable option for providing continuous communication with the back side of the moon.

No foreseeable schedule delays or budget changes are expected. The project will be completed on time and within budget by December 7, 2000, when the final report will be submitted to NASA.

ORGANIZATION OF A PROGRESS REPORT

The sample report uses standard sections to outline progress: Project Description, Work Accomplished, and Work Remaining. Here are the pieces of information each of these sections would typically cover or the questions they would answer:

Project Description

- purpose of project
- specific objectives
- scope (limitations, if any)
- alternative solutions being evaluated
- criteria used for evaluation
- dates project began and budget approved
- people working on project
- client (people for whom work is being done)
- other necessary background information

Work Accomplished

- What specific tasks have been accomplished?
- Does the progress of work actually match the schedule in the proposal?
- Which scheduled tasks are still outstanding?
- Which tasks are new?
- Make clear how you have dealt with problems and changes, if any. (Discuss more fully than in The Project Description.)

Work Remaining

- What specific tasks are left to do?
- How will you use the remaining tasks to make a recommendation to the client?
- What are the final deliverables (often a report or a presentation)?
- Is the project on schedule and within budget?

8.6 SAMPLE FEASIBILITY REPORT

This document reports on a study conducted for the owners of a set of apartment complexes in a medium-sized American city. Tenants have been complaining about noise from neighboring apartments, so the owners have hired a consulting engineering firm to investigate possible cost-effective means of controlling the transmission of sound between apartments. The student consultant has spent the semester defining the problem, defining a scope of work, and evaluating chosen solutions on the basis of a set of criteria worked out with the client. The report was produced for a Technical Communication class, but the methodology and analysis were very professional. And the client was real.

The organization of this report is typical of the way feasibility reports are put together: sections defining the problem at hand, describing how the list of feasible solutions was narrowed down, explaining the criteria used to measure the feasibility of various solutions, presenting the findings of the investigation, applying a method for final evaluation of feasibility, and finally recommending a course of action.

8.6.1 A Word about Using the Sample in This Section

This section presents a model of a good feasibility report for you to use. The model may not, however, fit the purpose or the requirements of a document you are called upon to write. Always ask your boss or instructor for samples to look at. Don't then copy them slavishly by using a fill-in-the-blanks approach, but rather notice the type of information presented in each document or section and decide whether your information fits that type.

FINAL REPORT

Evaluation of Wall Designs That Control Airborne Sound Transmission

Firm Ideas

FRONT MATTER: COVER PAGE

A report's *cover page* should be designed for ease of reading and simplicity of concept. Like the cover of a book, a report's cover should give some idea of what is inside, without presenting too much detail. Unlike a book's cover, the covers of most technical reports do not name an author, but rather include the logo and name of the company or agency that produced the report (in this case, Firm Ideas). This sample cover has just three pieces of information: the type of report, the title, and the name of the authoring company. Many agencies have coding and numbering systems that require certain numbers to be on the cover of every report; at the very least, you want to let the reader know what type of document he or she is looking at. For example, is it a proposal, a progress report, or the final report on a particular project?

The cover page in this example also has a graphic depicting the subject matter of the work reported, but many technical reports use only a logo on the cover, reserving graphical display for conveying the technical content inside the report. Check with your instructor or company before designing your own cover. Very often, you will simply use a standardized cover and insert your own title.

The title, of course, is critical to the reader's understanding of the scope of what is reported here. Make sure the title on the cover page is exactly the same as on the title page (which comes a little bit later). Be straightforward and direct in phrasing the title, but offer as much detail as a reader will need in order to understand the scope of the work reported. Compare this sample report's title,

> Evaluation of Wall Designs That Control Airborne Sound Transmission

with another version of it:

> Evaluation of Six Wall Designs That Are Intended to Control Airborne Sound Transmission between Apartments via a Common Wall

The second title probably offers more information than is necessary to understand the scope of work. On the other hand, the following title does not offer enough information about this study as done for a particular client with particular sound-control problems:

> Controlling Sound Transmission

The implied scope is much too large here: There are many ways other than designing a wall to control sound. And there are ways to transmit sound other than exclusively through the air. So we don't really understand the scope of this particular study from the title alone.

FIRM IDEAS

100 East 51st Street   Austin, TX 78705   (512) 400-3333

May 1, 2004

Mike Parker, Manager of EBC Properties
1000 Summit Lane
Austin, TX 78705

Dear Mr. Parker:

The attached technical report contains Firm Ideas' evaluation of wall designs that control airborne sound transmission. This report includes the descriptions of each wall design, the evaluation process, and the wall design that Firm Ideas recommends EBC Properties use in their new apartment complexes. We used the solution criteria established by EBC Properties and Firm Ideas to evaluate six wall designs:

- Single-Stud Wall with Resilient Channel
- Single-Stud Wall with Sound-Deadening Board
- Staggered-Stud Wall
- Staggered-Stud Wall with Sound-Deadening Board
- Double-Stud Wall
- Double-Stud Wall with Sound-Deadening Board

Firm Ideas recommends that EBC Properties use the Double-Stud Wall in their new apartment complexes. First, the Double-Stud Wall provides moderate performance with an STC value of 53 decibels. Second, the Double-Stud Wall has a relatively low material cost of $11.98 per linear foot and a relatively low labor cost of $8.74 per linear foot. Finally, the Double-Stud Wall can support the highest mountable object weight, 110 pounds per linear foot. This wall design best meets the needs of EBC Properties.

Firm Ideas is available to answer any questions you may have concerning this report or the project in general. You can contact Firm Ideas at (512) 400-3333 or by e-mail at *engineers@mail.utexas.edu.*

Sincerely,

Jeffrey Hunt
Project Engineer

FRONT MATTER: LETTER OF TRANSMITTAL

The *letter of transmittal* is a way of conveying the report to the right person: the decision maker who will act upon the work reported in the document. Like any letter, it is directed to a person, not a company or a department. Essentially, the letter gives a brief summary of the executive summary, answering the following questions: What was the scope of the work done, and why was it done? What was found? What conclusions and recommendations can be made on the basis of the work done? If a recommendation is made, how much will it cost to implement it? Sometimes the transmittal letter also addresses concerns that management might have with the results of the work—environmental concerns, especially, or possible legal consequences. The author may not have specifically studied the environmental consequences, for instance, of developing a piece of land, but may have become aware through developing a master plan that neighborhood groups are concerned about runoff to an aquifer.

Our sample letter does not convey any such concerns, but it offers specific conclusions about the cost and performance of the recommended wall design. Comparison with the other wall designs studied is left to the executive summary.

Sometimes, the transmittal letter is clipped onto the front of the report, so that the decision maker for whom it is primarily intended can remove it. That way, the letter remains more private, and the rest of the report can be photocopied and circulated to other readers without including remarks meant just for the primary reader. If your letter does not need to contain any such remarks, then binding the letter into the report makes sense so that they do not become separated.

Remember to sign your letters!

Evaluation of Wall Designs That Control Airborne Sound Transmission

Prepared for

EBC Properties

Prepared by

Jeff Hunt

Firm Ideas

May 1, 2004

FRONT MATTER: TITLE PAGE

This sample report has a title page as well as a cover page. Certainly, in many technical documents, the cover and title page are one and the same. Separating them, however, allows you to create a more visually prominent design for the front cover (see the graphics on this sample's cover) and to include less text on the cover. Again, check with your company or instructor to see which format is expected or required. If the date is not included on the front cover, it must definitely appear on the title page. Providing the date is an important way to establish your document's place in the continuum of work and research on the particular topic or problem.

Company or agency policy determines whether the author's name actually appears anywhere on a technical document. Oftentimes, the company or agency is considered to be the author, because in reality many reports are collaboratively produced. Chances are that as a working engineer, you will not research and write reports all by yourself. But you may be the project lead or have some other such position that requires you to take responsibility for the document.

This title page has a typical format: the client (in this case, only the company, not a person) and the author are listed after "Prepared for" and "Prepared by," respectively. If you look closely, you will notice that the client's name is in a slightly bigger font than the author's. Truly, the individual author's name is probably the least important identity on the title page.

Table of Contents

List of Illustrations

Figures

Tables

FRONT MATTER: TABLE OF CONTENTS AND LIST OF ILLUSTRATIONS

The *Table of Contents* (TOC) is more important than you might realize. Readers can glance at it and get a pretty good idea of the scope and subjects covered in the document. It is, after all, an outline, and thus it aids both the writer during the writing process and the reader during the reading process. But any TOC is only as good as its design and phrasing are clear and accurate, respectively. The headings must match the actual headings used in your report. But do not feel that you must include all levels of headings; two or three levels are usually sufficient.

Your word processor can help you create clear TOCs, so get to know the capabilities of yours. A program such as Microsoft Word can create an outline view that becomes the TOC, and thus, both the document's headings and the TOC can be updated at the same time. Do not use the space bar to try and insert the spaces in between a heading and the page number. Instead, set tabs or create a table so that the numbers will be lined up properly. The little dots connecting the words and page numbers in the TOC are called a "leader" (found under the Tabs function in Word). The dots do help the reader's eye make the visual connection.

If you have appendices, be sure to give each one a letter (A, B, C, etc.) and an informative title. We want to see at a glance what material is covered in an appendix; it is disconcerting to turn to an appendix and see a bunch of photocopied, handwritten, or printed pages and not have a clue as to what information they contain and how that information relates to your report. The sample report's Appendix A is called "Material and Labor Cost Calculations"; thus, we expect to see detailed cost figures and perhaps some formulas for arriving at the cost data included in the body of the report. Turn to the page following the actual appendix of this book for more information about developing, formatting, and referring to appendices.

A list of illustrations—or a list of figures and tables—is commonly used only if you have more than a total of five numbered graphics in your report.

Generally, neither the table of contents nor the list of illustrations is paginated.

ii   **Executive Summary**

Firm Ideas was hired by EBC Properties to evaluate wall designs that will control airborne sound transmission in their new apartment complexes. In their existing complexes, EBC Properties has received complaints of transmitted noise through the common wall between two apartments, and the high levels of noise have caused many residents to move out in search of apartments with better sound control. To avoid a similar problem in their new complexes, EBC Properties would like to know which wall design provides adequate sound control through the common wall.

For this report, Firm Ideas evaluated six wall designs that provide airborne sound control:

- Single-Stud Wall with Resilient Channel
- Single-Stud Wall with Sound-Deadening Board
- Staggered-Stud Wall
- Staggered-Stud Wall with Sound-Deadening Board
- Double-Stud Wall
- Double-Stud Wall with Sound-Deadening Board

Firm Ideas used four criteria to evaluate the wall designs. In order of importance, the criteria are: performance, material cost, labor cost, and mountable-object weight. Performance was evaluated by obtaining the Sound Transmission Class (STC) values for each wall design. Firm Ideas calculated the material cost and labor cost by using industry-standard cost estimation techniques. The mountable-object weight for each wall design was found in material specifications and on-line sources.

The wall design evaluation by Firm Ideas produced the following findings. All designs met the minimum STC value of 50 required by the *1997 Uniform Building Code—Section 1208*. The Double-Stud Wall with Sound-Deadening Board provided the best performance, with an STC of 58 decibels, and the Staggered-Stud Wall provided the worst performance, with an STC of 50 decibels. The material costs ranged from $10.53 per linear foot for the Staggered-Stud Wall to $16.38 per linear foot for the Double-Stud Wall with Sound-Deadening Board. The labor costs ranged from $6.93 per linear foot for the Staggered-Stud Wall to $13.02 per linear foot for the Double-Stud Wall with Sound-Deadening Board. Finally, the maximum mountable-object weight ranged from 50 pounds per linear foot for the Single-Stud Wall with Sound-Deadening iii

Board, Staggered-Stud Wall with Sound-Deadening Board, and Double-Stud Wall with Sound-Deadening Board to 110 pounds per linear foot for the Double-Stud Wall and Staggered-Stud Wall.

After having evaluated the six wall designs, Firm Ideas recommends that EBC Properties use the Double-Stud Wall as the common wall in their future complexes. This wall design provides moderate performance (STC value of 53 decibels), a relatively low material and labor cost ($11.98 per linear foot and $8.74 per linear foot, respectively), and the highest mountable-object weight. Our study reveals that this wall design is the best value for the money and gives tenants the best combination of benefits: increased sound control and a generous weight allowance for mounting cabinets and fixtures.

BODY OF THE REPORT: *EXECUTIVE SUMMARY*

The ability to summarize, both verbally and in writing, is one of the most important skills a technical professional can master. People want information in as selective and brief a form as possible. Behind every one- to two-page white paper produced for a congressperson headed into an appropriations meeting, there are weeks, if not months, of information gathering, research, and synthesis. Decision makers (whether lawmakers or division heads within an engineering company) must handle many projects and make decisions quickly. Thus, the term "executive summary" was born as a way of labeling for this audience the section of any document that will answer the big questions about time, cost, and resources required for a solution to the problem investigated.

The executive summary must move quickly through the most important information in the report: the reason the project was undertaken, the way in which the project was handled (methodology), the most important results, the conclusions, and any recommendations (along with their costs and benefits). All of this information is presented in much greater detail in the rest of the report, but remember that the executive audience may not read the rest of the report. Remember, too, that rather than photocopying the entire report, meeting organizers may choose to meet and make decisions based only on information in the executive summary, so the information must be accurate, brief, and self-contained. Do not refer to any page, graph, table, citation, or appendix in the rest of the report. The example executive summary here does refer to the *1997 Building Code*, however, because adhering to code is a prerequisite for any feasibility analysis.

Let us look more closely at the five paragraphs of this executive summary. The first paragraph outlines the client's problem and what remedies are being sought (the reason that Firm Ideas undertook the project). The problem with inadequate sound control in the apartment complexes will be described in much more detail in another section.

The second paragraph is a simple listing of the wall designs evaluated by Firm Ideas in their quest to find the most feasible alternative for the client, EBC Properties. How the engineers came up with these six alternatives does not need to be explained here; it will be described later.

The third paragraph sets out the criteria used to evaluate the six alternative wall designs. Since these criteria constitute the heart of this feasibility study, they are presented along with the sources from which information about them was obtained. Since this report happened to be for a technical-communication class, the decision maker (the instructor) wanted to know that the author had consulted reliable sources. In an industry report, such consultation would be assumed and could be listed later in a references section. The four criteria are described much more fully later.

The most important results of the analysis are presented in the fourth paragraph. Instead of giving major results for all six wall designs for all four criteria, the paragraph presents a range of results—enough data to support the conclusion and recommendation given in the final paragraph. Notice that the recommendation in the fifth paragraph is first given as a factual conclusion (the chosen wall design costs $11.98 per linear foot, provides "moderate" performance, etc.) and then stated in terms of its overall benefits. Most recommendations must be justified clearly in terms of cost and benefits.

This summary does not include details about the rating system used to assign points to each wall system. That level of explanation is left to the body of the report. Instead, the summary presents in shortened form a highly selective justification of the recommendation and answers the two questions most on the minds of an executive audience:

- What action is recommended?
- How much will that action cost to implement?

A good benchmark for length of an executive summary is 5–10% of the whole report. Some industry reports actually have very long executive summaries, because the authors are trying to stuff as much information as possible into the section that "executives" read. Be careful of this temptation.

Introduction

1

Background

EBC Properties, an Austin apartment management company, has been receiving complaints about transmitted sound through the common wall in their existing complexes. The high levels of noise have caused some residents to move out of EBC Properties's complexes in order to find apartments that provide better sound control. Meanwhile, EBC Properties is supervising the design of two new apartment complexes for the Austin area. To avoid similar sound control problems in their new complexes, EBC Properties has requested that Firm Ideas evaluate wall designs that provide adequate sound control through the common wall.

Purpose and Scope

To meet the needs of EBC Properties, Firm Ideas has evaluated wall designs that control airborne sound transmission, and we have recommended the best wall design. After EBC Properties reads this report, they plan to use the recommended wall design as the common wall in their new apartment complexes.

The new apartment complexes will be two stories high and constructed with wood framing. Therefore, Firm Ideas only evaluated wall designs framed with wood studs capable of supporting typical dead and live loads of apartments. Keeping these limitations in mind, we have evaluated six wall designs that provide airborne sound control:

- Single-Stud Wall with Resilient Channel
- Single-Stud Wall with Sound-Deadening Board
- Staggered-Stud Wall
- Staggered-Stud Wall with Sound-Deadening Board
- Double-Stud Wall
- Double-Stud Wall with Sound-Deadening Board

These six wall designs differ in their placement of studs (single, staggered, or double row of studs) and their sound-controlling materials (resilient channel and sound-deadening board).

2 **Methodology**

After choosing the six wall designs for evaluation in this study, Firm Ideas obtained performance data for each wall design. Performance information was obtained from *Construction: Principles, Materials, and Methods*, a building-construction textbook (L. Simmons, 2001). This textbook listed Sound Transmission Class (STC) values (a measure of performance) for each design.

Next, Firm Ideas calculated each design's material cost and labor cost. The costs were calculated using RS Means's *2002 Building Construction Cost Data*, an industry-standard cost estimation book.

Then, Firm Ideas estimated the maximum mountable-object weight for each wall design. These maximum weights were obtained from manufacturer specifications. In addition, we estimated the mountable-object weights for some designs by using recommendations from on-line construction articles.

Finally, we used a point system to determine the best wall design. This point system enabled us to combine the results from our evaluation with the solution criteria to determine the best wall design. A detailed explanation of the point system is given in the Factual Summary and Rating Procedure section.

Organization

The rest of this report is divided into sections that give detailed information on the alternative wall designs, the criteria used to evaluate each wall design, and the evaluation process used to make a recommendation. The first section, Wall Design Alternatives, contains a general discussion on the basics of sound control and includes descriptions of each wall design. The second section, Solution Criteria, describes the solution criteria used to evaluate each wall design, and the third section, Evaluation of Wall Designs, provides an evaluation of each wall design. The evaluations in this section are divided into subsections that compare the wall designs according to the solution criteria. The fourth section, Factual Summary and Rating Procedure, summarizes the comparisons made in the third section and gives the step-by-step procedure that Firm Ideas used to make a recommendation. The Conclusions section presents the findings from the rating procedure, and in the last section, Firm Ideas recommends the best wall design for EBC Properties to use in their new apartment complexes.

BODY OF THE REPORT: *INTRODUCTION*

The introduction to any report must orient the reader enough that he or she can understand quickly the more detailed and technical information contained in the report's central sections. Like the executive summary, the introduction should be written for the greatest possible number of readers—semitechnical and nontechnical as well as technical. How do you orient readers? For client-focused reports (as opposed to journal articles), you make sure to answer these questions:

1. What is the problem or need that necessitated the work reported here?
2. How did the report's author become involved with this problem or need?
3. How did the authors go about attacking the problem, including defining the scope of work?
4. For a recommendation report, how did the author evaluate solutions to the problem?
5. Where can the reader find particular types of information in this report?

The subheadings here reflect the attempt to answer each of these questions. The "Background" subsection answers the first and second questions, "Purpose and Scope" addresses the third question, "Methodology" the fourth, and "Organization" the fifth. Many introductions do not address the fifth question, figuring that a reader can always look at the table of contents. Indeed, if all the "Organization" subsection does is repeat the report's headings, then the section is not necessary. It might be useful, however, to illustrate here what the rating procedure is or explain why the "Evaluation" section is broken down by criteria.

Notice that this section, like all sections in this report, makes liberal use of headings and lists to chunk similar information together and to allow the reader to find particular information quickly.

Wall Design Alternatives   3

After a brief description of the creation and control of airborne sound transmission, this section describes the wall designs evaluated by Firm Ideas for use in EBC apartment complexes.

Basics of Sound Control

Airborne sound is created from sources such as voices, televisions, and radios (McMullan, 1991). A source sound in one room puts sound energy into the air in the form of sound waves. When these sound waves contact a wall, the waves in the air are converted to vibrations in the wall structure. Then, the vibrations travel through the wall structure and the space in the wall cavity. When the vibrations emerge on the opposite side of the wall, the vibrations are converted back to sound waves, which create a sound in the adjoining room.

Airborne sound transmission through walls can be controlled using three methods:

1. Increasing the weight of the wall: Heavy and dense materials provide good sound control because the sound energy cannot easily vibrate heavy materials.

2. Using discontinuous wall framing: Separating the framing members provides good sound control because the sound vibrations cannot follow a direct path through the wall structure to the other side of the wall.

3. Placing sound-isolating materials in the wall: Adding sound-isolating materials in the wall provides good sound control because the flexible nature and fibrous structure of these materials isolate and absorb sound energy.

Wall Designs

Firm Ideas has evaluated six wall designs (common in apartment complexes) that use all three methods of sound control:

- Single-Stud Wall with Resilient Channel
- Single-Stud Wall with Sound-Deadening Board
- Staggered-Stud Wall
- Staggered-Stud Wall with Sound-Deadening Board
- Double-Stud Wall
- Double-Stud Wall with Sound-Deadening Board

The descriptions of the wall designs are given on the next two pages.

4

Single-Stud Wall with Resilient Channel

This wall design is framed with a single row of 2 × 4 studs spaced 16" apart along a 2 × 4 wood plate. A $\frac{5}{8}$"-thick gypsum board is nailed to one side of the wall, and the gypsum board on the other side is attached to the wood studs by a resilient metal channel. The dense gypsum board provides increased weight to the wall, and the resilient channel acts as a spring to isolate and absorb the vibrations in the studs. The absorption by the channel reduces the vibrations in adjoining gypsum board, thus reducing the sound transmission to the other side of the room. The fiberglass insulation in the wall cavity absorbs additional sound energy (McMullan, 1991; Simmons, 2001).

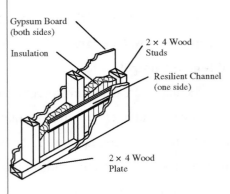

Single-Stud Wall with Sound-Deadening Board

This wall design is framed with a single row of 2 × 4 studs spaced 16" apart along a 2 × 4 wood plate. A ½"-thick sound-deadening board is adhesively applied to each side of the studs. Then, a $\frac{5}{8}$"-thick gypsum board is adhesively applied to the faces of the sound-deadening board. The dense gypsum board provides increased weight to the wall, and the sound-deadening board acts as an isolating material by absorbing vibrations in the studs. This vibration absorption by the sound-deadening board reduces the vibration sent to the opposite gypsum board. The fiberglass insulation in the wall cavity absorbs additional sound energy (McMullan, 1991; Simmons, 2001).

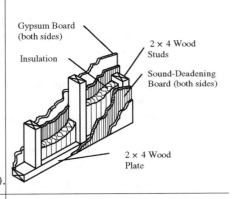

Staggered-Stud Wall

This wall design is framed with a single row of 2 × 4 studs staggered along a 2 × 6 wood plate. The staggered studs are placed every 8", so each $\frac{5}{8}$"-thick gypsum board is nailed to a stud every 16". The dense gypsum board provides increased weight to the wall, and the staggered pattern allows each gypsum board to be nailed to its own set of studs. This discontinuous construction reduces the direct transmission of sound vibration from one side of the wall to the other side of the wall. The fiberglass insulation in the wall cavity absorbs additional sound energy (McMullan, 1991; Simmons, 2001).

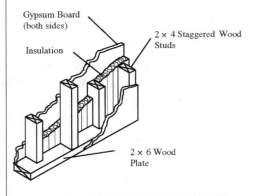

Staggered-Stud Wall with Sound-Deadening Board

This wall design is framed exactly the same as the *Staggered-Stud Wall*. However, ½"-thick sound-deadening board is adhesively applied to each side of the studs. Then, $\frac{5}{8}$"-thick gypsum board is adhesively applied to the faces of the sound-deadening board to increase the weight of the wall. The staggered studs provide resistance to sound transmission just as in the *Staggered-Stud Wall*, but the added sound-deadening board absorbs even more sound energy that may travel through the 2 × 6 wood plate or wall cavity. The fiberglass insulation in the wall cavity absorbs additional sound energy (McMullan, 1991; Simmons, 2001).

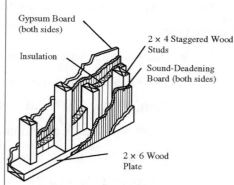

Double-Stud Wall

This wall design is framed with two rows of 2 × 4 wood studs placed on separate 2 × 4 wood plates. The studs in each row are spaced 16" along the plates. A $\frac{5}{8}$"-thick gypsum board is nailed to the outer face of each row of studs to increase the weight of the wall. Because the two rows of studs are completely unattached, the sound waves that vibrate one gypsum board do not directly move through the studs to the opposite gypsum board. This discontinuous construction reduces the direct transmission of sound vibration from one side of the wall to the other side of the wall. The insulation in the wall cavity absorbs additional sound energy (McMullan, 1991; Simmons, 2001).

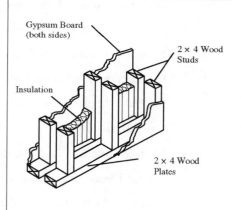

Double-Stud Wall with Sound-Deadening Board

This wall design is framed exactly the same as the *Double-Stud Wall*. However, ½"-thick sound-deadening board is adhesively applied to the outer face of each row of studs. Then, $\frac{5}{8}$"-thick gypsum board is adhesively applied to the faces of the sound-deadening board to increase the weight of the wall. The separated rows of studs provide resistance to sound transmission just as in the *Double-Stud Wall*, but the added sound-deadening board absorbs even more sound energy that may travel through the wall cavity. The insulation in the wall cavity absorbs additional sound energy (McMullan, 1991; Simmons, 2001).

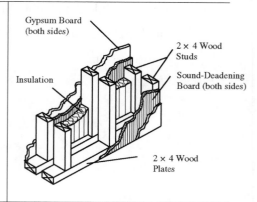

Source for all drawings: Simmons, Leslie. *Construction: Principles, Materials, and Methods.* 2001.

6

Summary of Wall Designs

The similarities and differences among the wall designs are given as follows (grouped by the framing type):

- Single-Stud Walls: Both the Single-Stud Wall with Resilient Channel and the Single-Stud Wall with Sound-Deadening Board are constructed with single-stud framing, gypsum board, and fiberglass insulation. However, the Single-Stud Wall with Resilient Channel uses a resilient metal channel to reduce the vibrations from airborne sound energy, while the Single-Stud Wall with Sound-Deadening Board uses sound-deadening board to reduce the vibrations.

- Staggered-Stud Walls: Both the Staggered-Stud Wall and the Staggered-Stud Wall with Sound-Deadening Board are constructed with staggered-stud framing, gypsum board, and fiberglass insulation. However, the Staggered-Stud Wall uses only the staggered studs to reduce the vibration transmission from airborne sound energy, while the Staggered-Stud Wall with Sound-Deadening Board uses both the staggered studs and sound-deadening board to reduce the vibration transmission.

- Double-Stud Walls: Both the Double-Stud Wall and the Double-Stud Wall with Sound-Deadening Board are constructed with double-stud framing, gypsum board, and fiberglass insulation. However, the Double-Stud Wall uses only a double row of studs to reduce the vibration transmission from airborne sound energy, while the Double-Stud Wall with Sound-Deadening Board uses both the double row of studs and sound-deadening board to reduce the vibration transmission.

BODY OF THE REPORT: THE CENTRAL SECTIONS

Wall Design Alternatives

The "Wall Design Alternatives" section presents and describes the wall systems that Firm Ideas chose as worthy of a more in-depth evaluation of sound-control effectiveness. This section expands upon the simple listing of alternatives given in the Executive Summary and Introduction. Here, the alternative walls are described in words and pictured in sketches. A departure from the norm is the treatment of illustrations as part of the text rather than as figures. The sketches have titles, but no figure numbers. The two pages of sketches look more like part of an instruction manual than a technical report. The decision not to number the figures was made because standard numbering and titling of six different figures would add unnecessary detail to a clear presentation of the alternatives. Before you make a similar decision, check with your boss or instructor. Notice that deciding not to number the sketches does *not* mean that you shouldn't cite their source; see the "Source for all drawings" citation at the bottom of the last sketch.

Like many of the initial sections of a technical report, this section provides some background on the technical concepts involved in the work reported. Airborne sound transmission and the most common methods of controlling it are not subjects that a real-estate development company would necessarily know much about, so the author provides some background. On the other hand, although whole books have been written on the subject, the client does not need to be educated on the full technical details of sound transmission or the nuances of controlling sound. An overview is sufficient here.

Likewise, at the end of this section, the author gives the reader a comparative summary of the wall designs presented separately on previous pages. Such comparative analyses are especially useful in evaluation reports and enable the reader to understand quickly the similarities and differences that will affect the ultimate recommendation. Notice the grouping of wall designs based on framing type, so that six wall designs end up in three bullet points. Chinking information like this helps the reader understand and remember concepts and facts.

Solution Criteria 7

Firm Ideas evaluated each wall design by using four solution criteria. EBC Properties set the performance, material cost, and labor cost criteria, and Firm Ideas added the mountable-object weight criterion. After the solution criteria were set, EBC Properties told Firm Ideas how much each criterion should influence the recommendation. The four solution criteria are given as follows in prioritized order, along with each criterion's influence on the recommendation:

1. **Performance**—constitutes 45% of the recommendation.

   Because the purpose of this project is to provide adequate sound control, EBC Properties stated that the performance of the wall design was the most important criterion. In this evaluation, performance was defined as how well each wall resisted airborne sound transmission. Therefore, Firm Ideas used the wall assembly's Sound Transmission Class (STC) rating to determine its sound transmission resistance. The STC values were obtained from *Construction: Principles, Materials, and Methods*, a building construction textbook (Simmons, 2001). An STC rating reflects the number of decibels a sound loses while passing through a given wall design; a better performing design produces a higher STC rating. All wall designs evaluated in this study have met the minimum STC rating of 50 decibels required by the *1997 Uniform Building Code—Section 1208*.

2. **Material Cost**—constitutes 30% of the recommendation.

   Like most apartment management companies, EBC Properties operates on a limited budget. Consequently, EBC Properties gave significant weight to the material cost of the wall designs. The material cost for each design was determined by the amount and type of the materials in that particular wall. Firm Ideas calculated the material cost by using RS Means's *2002 Building Construction Cost Data*, an industry-standard cost estimation book. This cost was measured in dollars per linear foot for an eight-foot-high wall.

8

3. **Labor Cost**—constitutes 15% of the recommendation.

The next criterion set by EBC Properties was the labor cost of the wall designs. The labor cost was defined as how much each wall costs to install, based on the amount and type of labor needed to construct the wall in the city of Austin. (The labor cost in the city of Austin is approximately 27% cheaper than the national average labor cost.) Firm Ideas calculated the labor cost by using RS Means's *2002 Building Construction Cost Data*. The labor cost was measured in dollars per linear foot for an eight-foot-high wall.

4. **Mountable-Object Weight**—constitutes 10% of the recommendation.

After construction is complete, the residents may use the common wall for mounting objects such as cabinets and fixtures; however, these objects affect the sound performance of some wall designs (USG, 2003). The mountable-object weight was defined as the maximum mountable weight each wall design can support without a decrease in sound performance. Firm Ideas obtained the maximum mountable weight from manufacturer specifications and recommendations from on-line construction articles.

BODY OF THE REPORT: THE CENTRAL SECTIONS

Solution Criteria

For a feasibility report, the "Solution Criteria" section constitutes the heart of the evaluation of various solutions to the problem or need being investigated. Only by setting valid, useful criteria can an investigator make technically correct and economically and socially feasible choices. Thus, defining, justifying, and prioritizing these criteria becomes critical in a feasibility investigation. This report not only prioritizes the evaluation criteria, but also provides a scale by which to measure exactly how important each criterion is to the final decision. Notice that the author is careful to point out that these criteria and their relative importance have been agreed to by the client.

If criteria are going to be used to measure the feasibility of various solutions, they must themselves be able to be measured. If cost, for instance, is a criterion for choosing something—a birthday present for your mother, let us say—then you have to know the costs of various possible presents. And those costs must all be in a common currency—U.S. dollars, let us say. In addition, you need to look at possible hidden costs of the candidate items—warranties or maintenance, for example. Notice that for this report on sound-controlling wall systems, cost is separated into material cost and labor cost. Both those costs are measured in dollars per linear foot, but the first is calculated on the basis of the amount and type of material and the second on the time to construct. In addition, performance is specifically defined as the ability to resist airborne sound transmission, measured by the wall's STC rating.

For all criteria, the sources of information that were used for comparative measurement are specifically cited in parentheses at the end of the sentence containing the fact, finding, or quotation from that source. Different engineering journals and professional societies use somewhat different referencing and citation systems. In some systems, a superscript number is used at the end of the sentence. This number refers to an item in a list at the end of the report and signals that the sources are listed in the order they were first cited in the text. In other referencing systems, sources are cited in the text by author and year and are listed at the end of the report in alphabetical order. Since this report is on a civil- or architectural-engineering subject, the author uses the citation and referencing system required by ASCE (American Society of Civil Engineering), which uses author–date citation and an alphabetized list at the end of the report. Each branch of engineering has its own requirements for referencing, which can usually be found on the branch's website. See the ASCE requirements at http://www.pubs.asce.org/authors/index.html#ref.

Evaluation of Wall Designs   9

Firm Ideas evaluated six wall designs, using the four solution criteria established by EBC Properties and Firm Ideas. The six wall designs are given as follows, along with their abbreviations used in this section:

- Single-Stud Wall with Resilient Channel   (SSWw/RC)
- Single-Stud Wall with Sound-Deadening Board   (SSWw/SDB)
- Staggered-Stud Wall   (StSW)
- Staggered-Stud Wall with Sound-Deadening Board   (StSWw/SDB)
- Double-Stud Wall   (DSW)
- Double-Stud Wall with Sound-Deadening Board   (DSWw/SDB)

The following subsections (organized by the solution criteria) contain the results from our evaluation.

Performance

According to Simmons (2001), the DSWw/SDB has the highest STC value of 58 decibels (dB) and therefore has the best performance. The StSWw/SDB, DSW, and SSWw/SDB provide moderate performance, with STC values of 54, 53, and 52 dB, respectively. Finally, the SSWw/RC and the StSW have the worst performance, with STC values of 51 and 50 dB, respectively. These STC values are shown in Figure 1.

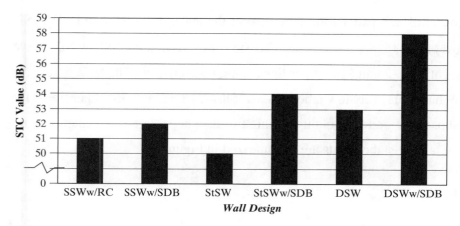

Figure 1. Performance: STC Values for Each Wall Design.
Source of data: Simmons, L., 2001. *Construction: Principles, Materials, and Methods.*

10 Material Cost

According to our calculations, the DSWw/SDB has the highest material cost of $16.38 per linear foot. The StSWw/SDB and the SSWw/SDB have moderate material costs of $14.93 and $13.35 per linear foot, respectively. The SSWw/RC, the DSW, and the StSW have relatively low material costs of $11.72, $11.98, and $10.53 per linear foot, respectively. All material costs are for an eight-foot-high wall. The material cost calculations we developed are shown in Appendix A, and the material costs are summarized in Figure 2.

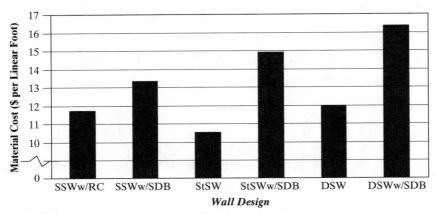

Figure 2. Material Cost per Linear Foot for Each Eight-Foot-High Wall Design.
Cost computed using RS Mean's *2002 Building Construction Cost Data.*

Labor Cost

The labor costs follow a pattern similar to that of the material costs. The DSWw/SDB has the highest labor cost of $13.02 per linear foot. The StSWw/SDB and the SSWw/SDB have moderate labor costs of $11.22 and $10.34 per linear foot, respectively. The SSWw/RC, the DSW, and the StSW have relatively low labor costs of $8.74, $8.58, and $6.93 per linear foot, respectively. All labor costs are for an eight-foot-high wall. The labor cost calculations are given in Appendix A, and the labor costs are summarized in Figure 3.

11

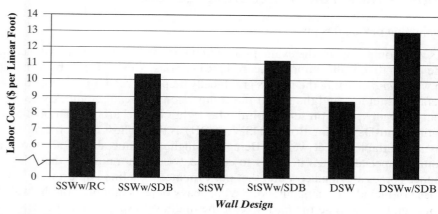

Figure 3. Labor Cost per Linear Foot for Each Eight-Foot-High Wall Design.
Cost computed with RS Means's *2002 Building Construction Cost Data*.

Mountable-Object Weight

According to Simmons (2001), the DSW and StSW have the highest mountable-object weight of 110 pounds per linear foot. The SSWw/RC has the next highest mountable-object weight of 67 pounds per square foot. The mountable-object weight for the SSWw/RC is less than for the DSW and StSW because a large amount of weight prevents the resilient channel from flexing and absorbing wall vibrations (Simmons, 2001). The SSWw/SDB, StSWw/SDB, and DSWw/SDB are the designs with the lowest mountable-object weights of 50 pounds per linear foot. These three designs have the lowest weight because when the sound-deadening board becomes overly compressed, the board loses its ability to flex and absorb wall vibrations (Georgia Pacific, 2003a). The mountable-object weights for each design are shown in Figure 4.

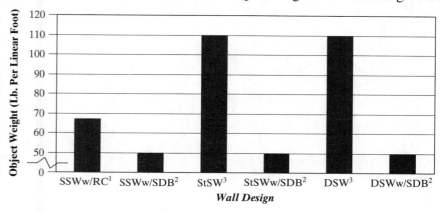

Figure 4. Maximum Mountable-Object Weight: Mountable-Object Weight (Pounds per Linear Foot) Each Wall Design Can Support.
Sources of data: 1. Handyman Club of America, 2003;
                             2. Georgia Pacific, 2003a;
                             3. Georgia Pacific, 2003b.

BODY OF THE REPORT: THE CENTRAL SECTIONS

Evaluation of Wall Designs

As you would suspect, the evaluation section contains a number of charts comparing the alternative wall solutions according to various criteria. Describing the findings in words *and* charting the results graphically helps the reader wade through a lot of detail and understand the gist of the comparisons. We can see quickly, for instance, that the wall system presented on the far right of the charts, the Double-Stud Wall with Sound-Deadening Board (DSWw/SDB), is the highest in all criteria except the last, mountable-object weight. Being the highest is not always a good thing, of course, especially when it comes to cost. So, we see in a nutshell that high performance comes at a high cost and that performance as defined in terms of STC rating does not include the capability of having heavy objects mounted on the wall. We do not even have to read the narrative of the findings to suspect that the DSWw/SDB will *not* be the recommended choice.

Notice that the sources of the *data* represented in each figure are listed underneath each figure's title. There is some disagreement on how detailed figure-data references need to be. If you have full publication information included in the list of references at the end of your report, you should not have to worry about providing details in the citation; for ASCE standards, the author and date should suffice. In citing the sources of cost estimates, however, this author felt that the full citation—RS Means's *2002 Building Construction Cost Data*—was a good idea.

Notice the break marks on the ordinate (the *y*-axis) of Figures 1–4. This mark is a responsible way of showing that the scale has been distorted. The interval from 0 to 50 decibels is a lot more compressed than the intervals between 50 and 51, 51 and 52, etc. Starting both axes at zero is a good idea, if you can possibly do it, but the values here all fall between 51 and 58, so there is no point in taking up a lot of room on the graph to show the scale below that range. Some graphing programs let you create a break mark, others do not. Even if you cannot make one, it is still a good idea to start both axes at zero.

12	**Factual Summary and Rating Procedure**

Factual Summary

After evaluating the six wall designs, Firm Ideas reached the factual conclusions shown in Tables 1–4. The designs are ordered from best to worst for each solution criterion. A narrative summary of these conclusions follows Tables 1–4.

Table 1. Wall Design Performance

Wall Design	STC (dB)	
Double-Stud Wall with Sound-Deadening Board	58	Best Design in Criterion
Staggered-Stud Wall with Sound-Deadening Board	54	
Double-Stud Wall	53	Decreasing Performance
Single-Stud Wall with Sound-Deadening Board	52	
Single-Stud Wall with Resilient Channel	51	
Staggered-Stud Wall	50	Worst Design in Criterion

Source: Simmons, Leslie. *Construction: Principles, Materials, and Methods.* 2001.

Table 2. Wall Design Material Cost

Wall Design	Material Cost ($ per Linear Ft)	
Staggered-Stud Wall	10.53	Best Design in Criterion
Single-Stud Wall with Resilient Channel	11.72	
Double-Stud Wall	11.98	Increasing Material Cost
Single-Stud Wall with Sound-Deadening Board	13.35	
Staggered-Stud Wall with Sound-Deadening Board	14.93	
Double-Stud Wall with Sound-Deadening Board	16.38	Worst Design in Criterion

Cost computed with RS Means's 2002 *Building Construction Cost Data.*

13

Table 3. Wall Design Labor Cost

Wall Design	Labor Cost ($ per Linear Ft)
Staggered-Stud Wall	6.93
Single-Stud Wall with Resilient Channel	8.58
Double-Stud Wall	8.74
Single-Stud Wall with Sound-Deadening Board	10.34
Staggered-Stud Wall with Sound-Deadening Board	11.22
Double-Stud Wall with Sound-Deadening Board	13.02

Best Design in Criterion

↓ Increasing Labor Cost

Worst Design in Criterion

Cost computed with RS Means's 2002 *Building Construction Cost Data.*

Table 4. Wall Design Maximum Mountable-Object Weight

Wall Design	Maximum Mountable-Object Weight (Lb per Linear Ft)
Double-Stud Wall	110
Staggered-Stud Wall	110
Single-Stud Wall with Resilient Channel	67
Single-Stud Wall with Sound-Deadening Board	50
Staggered-Stud Wall with Sound-Deadening Board	50
Double-Stud Wall with Sound-Deadening Board	50

Best Design in Criterion

↓ Decreasing Mountable-Object Weight

Worst Design in Criterion

Sources: Handyman Club of America; Georgia Pacific.

In the performance criterion, the Double-Stud Wall with Sound-Deadening Board was the best design, and the Staggered-Stud Wall was the worst design. Regarding material cost, the Staggered-Stud Wall was the best design, and the Double-Stud Wall with Sound-Deadening Board was the worst design. For labor cost, the Staggered-Stud Wall was the best design, and the Double-Stud Wall with Sound-Deadening Board was the worst design. In the mountable-object weight criterion, the Double-Stud Wall and Staggered-Stud Wall were the best designs, and the Single-Stud Wall with Sound-Deadening Board, Staggered-Stud Wall with Sound-Deadening Board, and Double-Stud Wall with Sound-Deadening Board were the least desirable designs.

14
Rating Procedure

After evaluating all the wall designs, Firm Ideas determined the best wall design, using a rating procedure. This procedure is outlined in the following steps:

1. Firm Ideas entered the information from each solution criterion into Tables 1–4 on the previous two pages. These tables order the designs from best to worst in each criterion.

2. For each solution criterion, we determined the number of points for each design:

    a. The best design received 10 points.

    b. The worst design or designs received 1 point each.

    c. The point values for the remaining designs were determined using linear interpolation. The linear interpolation for the performance criterion is given in Figure 5 as an example:

        i. A line was drawn from point (50,1) to point (58,10). This is the line in Figure 5.

        ii. A line was drawn from each design's STC value to the sloping line (resulting in the vertical arrows in Figure 5).

        iii. A line was drawn from the intersection of each vertical arrow and the sloping line to the *y*-axis (horizontal arrows). The point where the design's horizontal arrows intersect the *y*-axis is the design's point value.

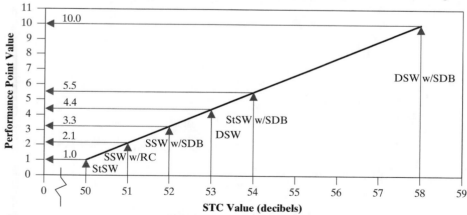

Figure 5. Example of Linear Interpolation—Performance. In this figure, the points are calculated for the performance criterion.

3. Using the point values determined from Step 2 and the solution-criteria weights determined by EBC Properties, Firm Ideas calculated the Total Weighted Points for each wall design as described in the "Conclusions" section.

BODY OF THE REPORT: THE CENTRAL SECTIONS

Factual Summary and Rating Procedure

Since the previous section presented so much numerical data, and because the alternative wall systems are hard to keep straight (whether as acronyms or long tongue-tying phrases), the author has included a "Factual Summary and Rating Procedure" section that gives the highlights of the findings in a visual form that makes misunderstanding impossible. As we noted in the last section, sometimes the "highest" bar in a bar chart means that the corresponding component performed well in that category, while other times it signifies the opposite. By showing how the wall systems are ordered from best to worst, each table presented in this section clarifies the comparisons for even the most inattentive reader.

The last subsection then describes the procedure whereby the wall systems were rated according to the importance of each category, or criterion. Linear interpolation is a common concept for engineers, but not for less technical people, so the interpolation is explained both in words and in Figure 5.

Together, these two subsections prepare us for the final sections, "Conclusions" and "Recommendation." We certainly know which wall "won" in the performance criterion, but the rating procedure prepares us to factor in a lot of other information before coming to conclusions.

Conclusions

15

For this study on sound-controlling wall systems, the six designs were compared according to four criteria and rated according to the procedure outlined in the previous section of this report. Firm Ideas calculated the Total Weighted Points for each wall design by using Equation (1).

$$\text{Total Weighted Points} = (PP \times 0.45) + (MCP \times 0.30) + (LCP \times 0.15) + (MOP \times 0.10) \quad (1)$$

PP is the Performance Point Value determined from Step 2;
MCP is the Material Cost Point Value determined from Step 2;
LCP is the Labor Cost Point Value determined from Step 2;
MOP is the Mountable-Object Weight Point Value determined from Step 2.

The calculations for the Total Weighted Points are shown in Table 5. Please see the tables in the "Factual Summary and Rating Procedure" for an explanation of the wall-system acronyms.

Table 5. Total-Weighted-Points Calculations Using Equation (1).

Wall Designs	PP	× Weight +	MCP	× Weight +	LCP	× Weight +	MOP	× Weight =	Total Weighted Points
SSWw/RC	1.0	× 0.45 +	8.2	× 0.30 +	7.6	× 0.15 +	3.6	× 0.10 =	**4.9**
SSWw/SDB	2.1	× 0.45 +	5.7	× 0.30 +	5.0	× 0.15 +	1.0	× 0.10 =	**4.0**
StSW	3.3	× 0.45 +	10.0	× 0.30 +	10.0	× 0.15 +	10.0	× 0.10 =	**6.0**
StSWw/SDB	4.4	× 0.45 +	3.2	× 0.30 +	3.7	× 0.15 +	1.0	× 0.10 =	**4.1**
DSW	5.5	× 0.45 +	7.8	× 0.30 +	7.3	× 0.15 +	10.0	× 0.10 =	**6.4**
DSWw/SDB	10.0	× 0.45 +	1.0	× 0.30 +	1.0	× 0.15 +	1.0	× 0.10 =	**5.1**

The Double-Stud Wall (DSW) had the most Total Weighted Points, and the Single-Stud Wall with Sound-Deadening Board (SSWw/SD) had the fewest Total Weighted Points. For a better visual comparison, the information in Table 5 is also given in Figure 6. This figure shows the breakdown of the Total Weighted Points for each wall design.

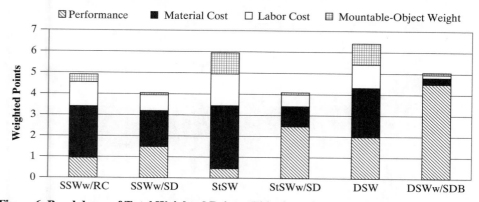

Figure 6. Breakdown of Total Weighted Points. This chart shows the breakdown of the total weighted points by solution criterion for each wall design.

BODY OF THE REPORT: CONCLUSIONS

At first glance, you might think that a conclusions section is unnecessary, given the factual summary presented just previously. Conclusions go further, however, than to simply present factual findings; they interpret and judge those findings. In this report, the interpretation is quantified by a rating system that applied a certain level of importance to each criterion and then assigned points to each wall system on the basis of relative importance. The point system is designed to give a total picture of how each wall rates across the board, including all categories. The interpretation is first given numerically (the equation and Table 5), then in narrative form (X is best, and Y is worst), and then in visual terms (Figure 6). The equation used to calculate the Total Weighted Points is explained clearly and should present no problem for any high school–educated reader. Figure 6 includes a caption ("This chart shows …") as a way of stressing quickly for all readers what is significant about the information shown on the chart. Like the executive summary and the introduction, conclusions should be written with a wide audience in mind.

Often, the conclusions are given along with the recommendation in a section called "Conclusions and Recommendations." Certainly, in this report, such a combined section would make sense. But separating the recommendation out allows for a greater focus on exactly what action is being recommended and avoids the tendency to weaken the recommendation by hiding it in the middle of a section or by using lots of passive voice.

16   **Recommendation**

After evaluating the six wall designs and following the recommendation procedure, Firm Ideas recommends that EBC Properties use the Double-Stud Wall as the common wall in their new complexes. This wall design provides moderate performance (STC value of 53 decibels) at a relatively low material cost and labor cost ($11.98 per linear foot and $8.74 per linear foot, respectively, for an eight-foot-high wall). This cost was the third lowest out of the six wall designs investigated. In addition, the Double-Stud Wall can support the highest mountable-object weight: 110 pounds per linear foot. Our evaluation reveals that this wall design is the best value for EBC Properties and gives their tenants the best combination of benefits: increased sound control and a generous weight allowance for mounting cabinets and fixtures. This wall design best meets the needs of EBC Properties.

BODY OF THE REPORT: RECOMMENDATION

The recommendation answers the client's most important questions: What should be done to solve the problem, and how much will it cost to do it? The recommendation should reveal good knowledge of the client's concerns, and the language should be straightforward and firm. If you do not feel quite sure about your recommended action, you should not make a recommendation, but rather say that further study is needed or that none of the solutions is feasible at present. Sometimes, recommendations can be broken down into short-term (sometimes called Band-Aid® solutions) and long-term recommendations. The implementation of long-term solutions often has to wait for adequate funding.

List of References   17

Bergman, P. (2002). "Soundproofing." *Ultra High Fidelity Magazine*. Issue 63.
   http://www.uhfmag.com/Issue63/soundproofing.html (10 March 2003).

Georgia-Pacific Co. "Georgia-Pacific Tough Rock Sound Deadening Board."
   http://www.gp.com/gypsum/gypsumboard/submittal/sounddeadening.html (1 April 2003).

Georgia-Pacific Co. "Georgia-Pacific Tough Rock Gypsum Board."
   http://www.gp.com/gypsum/gypsumboard/submittal/gypsumboard.html (1 April 2003).

Handyman Club of America. "Quiet." http://www.handymanclub.com/document.asp (1 April 2003).

Harris, C. (1991). *Handbook of Acoustical Measurements and Noise Control*. 3d ed. New
   York: McGraw-Hill.

International Conference of Building Officials. *1997 Uniform Building Code*. Volume 1.
   Whitter, CA: IBCO.

McMullan, R. (1991). *Noise Control in Buildings*. Boston: BSP Professional.

RS Means. *2002 Building Construction Cost Data*. 60th ed. Kingston, MA: RS Means.

Simmons, L. (2001). *Construction: Principles, Materials, and Methods*. 7th ed. New York:
   John Wiley & Sons.

Thorburn Associates, Inc. "Resilient Channels: How Metal Absorbs Sound Waves."
   http://www.ta-inc.com/newsletter/rslchnls.htm (14 March 2003).

USG Manufacturers. "Sound Construction."
   http://www.usg.com/Design_Solutions/2_3_sound_construct.asp (3 March 2003).

BACK MATTER: LIST OF REFERENCES

The "List of References" section (sometimes called "Works Cited") provides evidence that you thoroughly researched your subject and made use of creditable sources of information. Not all engineering reports contain references. A structural design report, for instance, may justify the design it presents on the basis of the structure's and the site's constraints alone, without referring to other designs published other places. For most university courses and for many feasibility reports, however, you want to show that you have gathered and synthesized good information in the process of evaluating engineering solutions.

Why is this list called "List of References" rather than "Works Cited"? Well, the title "Works Cited" means exactly what it says: All the material included in the list is cited somewhere in the text of the report. Thus, for example, for the eighth item on the list—RS Means's *2002 Building Construction Cost Data*—there are numerous citations in the body of the report, the first one appearing in the "Methodology" subsection of the Introduction. On the other hand, naming the list "List of References" means that it is more of a bibliography; you consulted each and every source on the list, but you may not have cited every one in your text. Thus, the article by P. Bergman, "Soundproofing," in the on-line *Ultra High Fidelity Magazine*, is not cited specifically in the body of this report; apparently, the author read this piece for background information but did not use in his or her report any specific fact, finding, or quotation presented by the article. In fact, three of the sources listed here are not referred to in the text (Bergman, Harris, and Thorburn).

For electronic sources, notice the date in parentheses at the end of the citation. This is the date the report author accessed the web site. If the site does not list a date this access date will be the only date in the citation. Since web sites can came and go, this date may provide the only proof of a particular electronic publication's existence.

The style of referencing in this list follows the ASCE guidelines for authors, found on the ASCE website http://www.pubs.asce.org/authors/index.html. ASCE requirements make sense here, since this is an architectural- and civil-engineering report. Make sure to check with your boss or instructor about which style of referencing is conventionally used in your field, class, or company. Often, larger companies (such as IBM) will have their own style guides for everything from abbreviations to reference lists.

A-1   Appendix A: Material and Labor Cost Calculations

Table A-1. Material Cost Calculations

Wall Design	Material	Cost per Unit of Material (dollars)	×	Amount of Material per Linear Foot	×	Austin Location Factor	=	Material Cost per Linear Foot
Single-Stud Wall with Resilient Channel	$\frac{5}{8}$" gypsum board	0.22 per SF	×	16 SF	×	0.948	=	3.34
	2 × 4 Wood Studs (16" o.c.) + plates	3.20 per LF	×	1 LF	×	0.948	=	3.03
	3.5" Insulation	0.34 per SF	×	8 SF	×	0.948	=	2.58
	Resilient Channel	0.73 per LF	×	4 LF	×	0.948	=	2.77
						Total =		**11.72**
Single-Stud Wall with Sound-Deadening Board	$\frac{5}{8}$" gypsum board	0.22 per SF	×	16 SF	×	0.948	=	3.34
	2 × 4 Wood Studs (16" o.c.) + plates	3.20 per LF	×	1 LF	×	0.948	=	3.03
	3.5" Insulation	0.34 per SF	×	8 SF	×	0.948	=	2.58
	Sound D. Board	0.29 per SF	×	16 SF	×	0.948	=	4.40
						Total =		**13.35**
Staggered-Stud Wall	$\frac{5}{8}$" gypsum board	0.22 per SF	×	16 SF	×	0.948	=	3.34
	2 × 4 Wood Studs (8" o.c.) + plates	4.87 per LF	×	1 LF	×	0.948	=	4.62
	3.5" Insulation	0.34 per SF	×	8 SF	×	0.948	=	2.58
						Total =		**10.53**
Staggered-Stud Wall with Sound-Deadening Board	$\frac{5}{8}$" gypsum board	0.22 per SF	×	16 SF	×	0.948	=	3.34
	2 × 4 Wood Studs (8" o.c.) + plates	4.87 per LF	×	1 LF	×	0.948	=	4.62
	3.5" Insulation	0.34 per SF	×	8 SF	×	0.948	=	2.58
	Sound D. Board	0.29 per SF	×	16 SF	×	0.948	=	4.40
						Total =		**14.93**
Double-Stud Wall	$\frac{5}{8}$" gypsum board	0.22 per SF	×	16 SF	×	0.948	=	3.34
	2 × 4 Wood Studs (16" o.c.) + plates	6.40 per LF	×	1 LF	×	0.948	=	6.07
	3.5 Insulation	0.34 per SF	×	8 SF	×	0.948	=	2.58
						Total =		**11.98**
Double-Stud Wall with Sound-Deadening Board	$\frac{5}{8}$" gypsum board	0.22 per SF	×	16 SF	×	0.948	=	3.34
	2 × 4 Wood Studs (16" o.c.) + plates	6.40 per LF	×	1 LF	×	0.948	=	6.07
	3.5" Insulation	0.34 per SF	×	8 SF	×	0.948	=	2.58
	Sound D. Board	0.29 per SF	×	16 SF	×	0.948	=	4.40
						Total =		**16.38**

Table A-2. Labor Cost Calculations   A-2

Wall Design	Material	Cost per Unit of Material (dollars)	×	Amount of Material per Linear Foot	×	Austin Location Factor	=	Labor Cost per Linear Foot
Single-Stud Wall with Resilient Channel	$\frac{5}{8}"$ gypsum board	0.24 per SF	×	16 SF	×	0.67	=	2.57
	2 × 4 Wood Studs (16" o.c.) + plates	4.00 per LF	×	1 LF	×	0.67	=	2.68
	3.5" Insulation	0.15 per SF	×	8 SF	×	0.67	=	0.80
	Resilient Channel	0.94 per LF	×	4 LF	×	0.67	=	2.52
							Total =	**8.58**
Single-Stud Wall with Sound-Deadening Board	$\frac{5}{8}"$ gypsum board	0.24 per SF	×	16 SF	×	0.67	=	2.57
	2 × 4 Wood Studs (16" o.c.) + plates	4.00 per LF	×	1 LF	×	0.67	=	2.68
	3.5" Insulation	0.15 per SF	×	8 SF	×	0.67	=	0.80
	Sound D. Board	0.40 per SF	×	16 SF	×	0.67	=	4.29
							Total =	**10.34**
Staggered-Stud Wall	$\frac{5}{8}"$ gypsum board	0.24 per SF	×	16 SF	×	0.67	=	2.57
	2 × 4 Wood Studs (8" o.c.) + plates	5.30 per LF	×	1 LF	×	0.67	=	3.55
	3.5" Insulation	0.15 per SF	×	8 SF	×	0.67	=	0.80
							Total =	**6.93**
Staggered-Stud Wall with Sound-Deadening Board	$\frac{5}{8}"$ gypsum board	0.24 per SF	×	16 SF	×	0.67	=	2.57
	2 × 4 Wood Studs (8" o.c.) + plates	5.30 per LF	×	1 LF	×	0.67	=	3.55
	3.5" Insulation	0.15 per SF	×	8 SF	×	0.67	=	0.80
	Sound D. Board	0.40 per SF	×	16 SF	×	0.67	=	4.29
							Total =	**11.22**
Double-Stud Wall	$\frac{5}{8}"$ gypsum board	0.24 per SF	×	16 SF	×	0.67	=	2.57
	2 × 4 Wood Studs (16" o.c.) + plates	8.00 per LF	×	1 LF	×	0.67	=	5.36
	3.5" Insulation	0.15 per SF	×	8 SF	×	0.67	=	0.80
							Total =	**8.74**
Double-Stud Wall with Sound-Deadening Board	$\frac{5}{8}"$ gypsum board	0.24 per SF	×	16 SF	×	0.67	=	2.57
	2 × 4 Wood Studs (16" o.c.) + plates	8.00 per LF	×	1 LF	×	0.67	=	5.36
	3.5" Insulation	0.15 per SF	×	8 SF	×	0.67	=	0.80
	Sound D. Board	0.40 per SF	×	16 SF	×	0.67	=	4.29
							Total =	**13.02**

BACK MATTER: APPENDICES

How do you decide what information to put in an appendix, as opposed to the body of the report? Well, the best way to decide is to ask yourself these two questions:

- Is all of this information critical for the reader to have at this point (in the body of the report)?
- Can I summarize the information in the body of the report and then put the details in an appendix, where they will not interrupt the reader?

If the answer to the first question is "yes," you do not want to relegate the information to an appendix, even if the answer to the second question is "yes." If the answer to the first question is "no," then definitely go ahead and put the information in an appendix. You can almost always summarize information. If you cannot, then go back to the first question; if the answer is still "no," then just refer the reader to the appendix. In any case, always remember to direct the reader to each appendix at the appropriate place in the report.

Typical kinds of information suitable for an appendix include large tables; complex calculations (more than a couple of equations); large or complex schematics, plans, or drawings; large maps; and raw data of any kind. Group similar kinds of data (e.g., calculations, photographs, survey results, rules, and requirements) in one appendix, and give each appendix a letter (A, B, etc.) and a meaningful title that indicates what sort of information it contains. The appendix here has the helpful title "Appendix A: Material and Labor Cost Calculations."

Normally, appendices are not paginated as part of the body of the report; rather, each appendix is paginated separately: A-1, A-2, B-1, B-2, etc.

- Making the design and format of a document consistent enhances the document's readability. Conversely, inconsistent headings, illogical spacing, and awkward integration of graphics in the text all lead readers to misunderstand or outright ignore your writing.
- Study models of well-presented information, but never copy them slavishly. Notice the type of information presented in each section of the sample document, and then decide whether your information fits that type.

8.1. Take the last homework problem you did for any engineering or science class and imagine that you have to explain the problem and its solution in a letter to the Director of Human Resources at an engineering company. The director wants the letter because he wants to see whether you can communicate technical information in a readable style. Use any graphic(s) you want but remember that the director has a business rather than an engineering degree. Use business letter format.

8.2. Write a proposal to your instructor describing a plan to study improvements to the parking situation for students at your school. Research and ponder some likely alternatives for creating more parking spaces and/or better traffic flow in and around campus. Include these items in a proposal of 2–4 pages:

   (a) Clear statement of traffic problem and its causes
   (b) Presentation of a couple of possible solutions you will investigate
   (c) A plan for completing the research in time to make a recommendation to your instructor by the end of the semester. What specific tasks do you need to accomplish in order to define the problem (observation is an important task!) and research possible solutions. Your *final* report will include suggestions for how best to alleviate the traffic problem, but this proposal presents a plan of study, *not* a recommendation.
   (d) A schedule for completing the research and writing tasks. Assume that your instructor will want a final recommendation report.
   (e) A short list of your qualifications for doing this investigation.

8.3. Write a paragraph describing which type and size font you use for writing print documents of more than one page. Explain the reasons for your choices.

8.4. For your next writing assignment, fill in the Style Sheet Checklist in Figure 8.4 (ignore any items that do not apply to the assignment). Show the checklist to your instructor.

8.5. True or false? Full justification of text always looks more professional than a ragged right margin.

8.6. True or false? Letters and memos are generally single-spaced.

8.7. A report's letter of transmittal is written primarily to whom?

   The head of the department most affected by the report

   The person who commissioned the study reported in the document

   The CEO of the company who commissioned the report

8.8. The summary of a document is usually about what percentage of the length of the whole document?

25%

10%

33%

8.9. Which is the best reason for creating a separate report section for any recommendations?

to highlight the recommendations

to highlight the conclusions

to strengthen the content of the report

8.10. The style of referencing used in most engineering journals and reports is based on which *one* of the following sources:

The Modern Language Association

The American Chemical Society

The Chicago Manual of Style

The Society for Technical Communication

FURTHER READING

Chicago Manual of Style, 15th ed., 2003. University of Chicago Press: Chicago. (Also available in CD-ROM.) The Manual is also available online: http://www.chicagomanualofstyle.org/home.html. You can try the online version for a free trial, but eventually you must subscribe to use it.
This is the mothership of style manuals. This big volume contains more than you'll ever want to know about manuscript preparation, documentation, and the publishing process, but it will also have definitive answers for you on all those niggling little questions about style (do you write out the number twelve or use the Arabic numeral?). The sections on "Grammar and Usage" and "Punctuation" are a terrific reference tool for engineers. Most discipline-specific style manuals in engineering are based on the *Chicago Manual*. See the "Documentation" section for a clear, if lengthy, demonstration of the differences between APA and MLA styles. Since you may have been taught the latter in high school, you will want to acquaint yourself with APA, now that you are in a technical field.

Walker, J.R. and Taylor, T. 2006. *The Columbia Guide to Online Style,* second edition. Columbia University Press: New York.
This reference book organizes and categorizes the new standards for citing electronic documents. It also includes sections on how to create documents electronically for print publication and how to format documents for online publication.

Index